Two books in one, *Death & Sex* explores the two facts of life that have dominated our thoughts, fears, and dreams since the earliest days of human history.

Why do things die? In his lucid and concise investigation, Tyler Volk reveals how creatures—from suicidal bacteria to fish, trees, and humans—actually use death to enhance life. Volk weaves together autobiography, biology, Earth history, and fascinating research to show us how death is a natural part of evolution and how thoughts of our own mortality affect our everyday lives. In the process, he proves that understanding what some have called the ultimate taboo can enrich the celebration of life.

Praise for *Death* by Tyler Volk

"In *Death & Sex* two of my favorite thinkers and writers ruminate on two of my favorite subjects and turn up all manner of unexpected interconnections. The result is a splendidly entertaining, informative, and original piece of science writing."
—JOHN HORGAN, author of *The End of Science*
and *Rational Mysticism*

"What delicious writing and reading! I love this wise and funny big-little book."   —ERICA JONG, bestselling author of *Fear of Flying*
and *Seducing the Demon*

"In a mere 90 pages, Tyler Volk's book *Death* brilliantly depicts the biology and psychology of its subject, putting death in proper perspective as an integral component of the life cycle. I've read many insightful books about death, but if I were to recommend one book to help someone come to terms with death, this would be it."   —JEFF GREENBERG, Director of the Social Psychology
Program, University of Arizona

"Tyler Volk succeeds in translating everything about the natural world with generous poetic details, from tree-filled landscapes to star systems, into one or another version of death. We humans are by-products of carbon dioxide from dead photosynthesizers, yet Volk manages to make even this a fact well worth celebrating."
—BETSEY DYER, Professor of Biology, Wheaton College,
author of *A Field Guide to Bacteria*

"Tyler Volk's *Death* is spark to the tinder of insight."
—HOWARD BLOOM, author of *The Lucifer Principle: A Scientific Expedition into the Forces of History*

"This champagne cocktail of exploration and insight, not to forget the murkier passions of lust, or the despondency that comes from unrequited love, abandonment, or loss—what an inspired confection of two immensities, sex and death. I genuinely can't recall reading a more inspiring or entertaining book in years!"
—FRANK RYAN, author of *Virolution* and *The Forgotten Plague*

"Dorion Sagan and Tyler Volk show us sex is optional and death is necessary, turning the tables on our lusts and fears, our origins and endings, in a surprisingly enticing way."
—ADAM DANIEL STULBERG, *Poetic Interconnections*

"Eschewing taboos and transgressing disciplinary boundaries, this volume manages to be, at once, both playfully iconoclastic and technically informative. Indeed, it exhibits the very rare capacity to popularize, without 'selling out' or oversimplifying, an intellectually challenging analysis. Where else is one going to experience such from chance encounters with de Sade, Monty Python, Bashō, and Poincaré?"
—SIMON GLYNN, Professor of Philosophy, Florida Atlantic University

"In *Death & Sex*—two books in one—quotidian simplicities·are dissolved in the acid of evolutionary theory. Death turns out to be more complicated than to be or not to be; and sex is seen to be far more complicated than a tale about a man, a woman, and a garden snake. Together, they form a pair of insightful lessons in the application of Darwinian concepts." —ANDREW LIONEL BLAIS, author of *On the Plurality of Actual Worlds*

"A boisterous Siamese twin of a book which looks at the two sides of the same molecular process—that of sex and that of death—within the framework of life almost eternal. Enjoy, and know you are part of it." —CRISPIN TICKELL, Director of the Policy Foresight Programme, Oxford University, and former UK Ambassador to the United Nations

DEATH

DEATH

TYLER VOLK

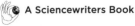 A Sciencewriters Book

CHELSEA GREEN PUBLISHING COMPANY
WHITE RIVER JUNCTION, VERMONT

A Sciencewriters Book

scientific knowledge through enchantment
Sciencewriters Books is an imprint of Chelsea Green Publishing. Founded
and codirected by Lynn Margulis and Dorion Sagan, Sciencewriters is an
educational partnership devoted to advancing science through enchantment in
the form of the finest possible books, videos, and other media.

Project Manager: Emily Foote
Developmental Editor: Jonathan Cobb
Copy Editor: Laura Jorstad
Proofreader: Helen Walden
Indexer: Christy Stroud
Designer: Peter Holm, Sterling Hill Productions
Cover Design: Kelly Blair

Printed in the United States of America
First printing September, 2009
10 9 8 7 6 5 4 3 2 1   09 10 11 12 13

Our Commitment to Green Publishing

Chelsea Green sees publishing as a tool for cultural change and ecological stewardship.
We strive to align our book manufacturing practices with our editorial mission and to
reduce the impact of our business enterprise in the environment. We print our books and
catalogs on chlorine-free recycled paper, using vegetable-based inks whenever possible.
This book may cost slightly more because we use recycled paper, and we hope you'll
agree that it's worth it. Chelsea Green is a member of the Green Press Initiative (www.
greenpressinitiative.org), a nonprofit coalition of publishers, manufacturers, and authors
working to protect the world's endangered forests and conserve natural resources. *Death
& Sex* was printed on Natures Natural, a 30-percent postconsumer recycled paper supplied
by Thomson-Shore.

Library of Congress Cataloging-in-Publication Data
Volk, Tyler.
 Death & sex / Tyler Volk and Dorion Sagan.
   p. cm.
 Includes bibliographical references and index.
 ISBN 978-1-60358-143-1
 1. Death. 2. Aging. 3. Sex. I. Sagan, Dorion, 1959- II. Title. III. Title: Death and sex.

 QP87.V647 2009
 612.6'7--dc22

                              2009030141

Chelsea Green Publishing Company
Post Office Box 428
White River Junction, VT 05001
(802) 295-6300
www.chelseagreen.com

All individual things pass away.
Seek your liberation with diligence.
—BUDDHA (sixth–fifth centuries BCE),
upon his deathbed

If you should have the desire to study Zen under a teacher and see into your own nature, you should first investigate the word shi [death]. If you want to know how to investigate this word, then at all times while walking, standing, sitting, or reclining, without despising activity, without being caught up in quietude, merely investigate the koan: "After you are dead and cremated, where has the main character [chief actor] gone?" Then in a night or two or at most a few days, you will obtain the decisive and ultimate joy.

—Japanese Zen Master HAKUIN (1686–1769)

Six times now I have looked death in the face. And six times Death has averted his gaze and let me pass. Eventually, of course, Death will claim me—as he does each of us. It's only a question of when. And how.

I've learned much from our confrontations—especially about the beauty and sweet poignancy of life, about the preciousness of friends and family, and about the transforming power of love. In fact, almost dying is such a positive, character-building experience that I'd recommend it to everybody—except, of course, for the irreducible and essential element of risk.

—Scientist and author CARL SAGAN (1934–1996)

CONTENTS

PART ONE

Connectivity

Impermanence

Is this what it's like to die?" I wondered. "Perhaps I really *am* dying."

During the winter of 1996–1997, I had moved from New York City to a secluded place where I could concentrate on writing. The book in progress was about life and Earth, so a trailer perched a mile up in the mountains of a spectacular and remote corner of New Mexico seemed ideal. But the metal container nearly became my coffin.

The first signs seemed innocuous. The tip of my right thumb went numb. Then at odd moments electrical zings began shooting along my arm. A few weeks later I started waking at night with painful cramps in one hand or the other. Once I was jolted awake to find my toes in contorted positions and half my face feeling like a wooden mask. Next, my hands and feet started "falling asleep" in the middle of the day and would not wake.

Medical care was a problem. My regular New York doctor was thousands of miles away. The nearest town was across two mountain ranges, and its sole neurologist flew in but once a week, weather permitting. So initially I hoped that my troubles would just go away on their own. Then I happened upon what I thought must be the ultimate cause of my infirmities: poisoning from carbon monoxide, emitted from a wall-mounted propane oven that had been activated just that winter after years of disuse. With the help of a meter I purchased, I discovered that airborne molecules of the odorless, invisible, deadly gas had at times been accumulating halfway to levels that could cause death in four hours. While writing about the atmosphere's CO_2, I ironically had been oblivious to my growing exposure to a related airborne gas whose biochemical lethality derived from one less oxygen atom.

I immediately shut down the oven, of course. Yet even so, to my horror I kept having what my neurologist over the phone termed "relapses." I grew more and more terrified as these "relapses" intensified. Soon I was barely able to write legibly. At night I found my mind trapped uncontrollably in inane obsessions. I imagined myself, for example, peeling an apple for hours, unable to cease or think of anything else. Coordination faltering, I had to steady myself when walking, one small step at a time. My chest would sporadically become the radiating center of body-filling pulsations, an uncontrollable drumming of rapid-fire vibratos that coursed along my arms and legs. Heartbeats pounded in my ears and set off reverberations all along my nerves.

Fearing that I could be fatally ill, I took to the outdoors and tried to make peace with myself during slow, clumsy walks in the valley that sheltered the trailer. On one cold, evening amble, with snow glossing the juniper trees and the shadows thickening, I relived my childhood and the ensuing pageant of my then forty-six years, trying to come to terms with my inner terror and the realization that, no matter what was going on, no guarantees had ever been given that I'd live to the standard life expectancy.

Over the following month I suffered several more "relapses," and my despair increased. Then one morning I startled myself with a new possibility: Could it be the old car?

I put the carbon monoxide meter in the front passenger's seat, started the engine, turned on the heating fan, and watched safely from the outside as the numbers surged into the danger zone. An exhaust leak! With every three-hour round trip to town I had been dosing myself with a second, independent source of carbon monoxide. I had come to the mountains for fresh air, but had found myself being poisoned twice over by defective technologies.

For more than a decade afterward I had to take an anticonvulsant drug to soothe what the doctors called "sensory distortions." Eventually I was able to wean myself from the medicine, apparently healed. But my outlook on life had permanently shifted.

During the time of terror, during the evening walks, I found myself taking refuge, even embracing, a deep core of gratitude. How marvelous to have lived at all, I felt. Had the carbon atoms of my body been locked into, say, the calcium carbonate crystals of limestone rock, then the atomic arrangements would have had more permanence. In that case, what about an "I"? The transient configuring of carbon in my body allows a conscious self to exist: complicated and conflicted, to be sure, yet also joyous, curious, and loving.

Sure, death would come. Death, I came to realize, was inherent both in my humanness and in the evolutionary nature of our existence. Life and death were totally intertwined. Life, a flowering of the fortunate way my atoms were combined, was bound up with inevitable death. In fact, death made life possible.

I have written this book not to instruct on how to reform and live right, however. Instead, I aim to explore how intertwined death is with life at every scale of our biology, our evolution, and our experience. I am not a healer or a public health worker, a hospice helper or a spiritual teacher. I am a natural scientist, a professional student of Earth's carbon cycle and lover of the patterns of all systems, from atoms to cognition.

The universe is an enormous field for any number of journeys into the dynamics of existence. Everywhere, at every scale from matter to mind, life and death appear woven together. Knowledge of these weavings, I believe, can nurture and deepen our daily lived experience of being.

⤎ 2 ⤏

Epicurean Dance of Friends

Epicurus (341–270 BCE) was among the most influential of ancient Greek philosophers. Though less widely known to the general public than Plato, Epicurus is cited by those who study cultural evolution of ideas as probably equal in importance. In America's history, Thomas Jefferson, who thought of himself as an Epicurean, is believed to have imported Epicurus's stress on happiness into the famous phrase in the Declaration of Independence that asserts the "pursuit of happiness" as an inalienable right.

Radical for his time, Epicurus invited women and slaves into the society of friends who gathered in his philosophical "Garden," probably some sort of commune in which participants grew their own vegetables and supported one another in the quest for happiness and pleasure. Some in later generations skewered the Epicureans as simple pleasure lovers—hedonists—above all else.

But students of the real Epicureanism know this philosophical school as one that has for good reason been likened to Buddhism, which had itself been born just a few centuries earlier. Both traditions emphasize a simple happiness based on the control of our desires. For example, indulging in occasional luxurious gluttony is perfectly fine, according to Epicurus, but one should not become dependent on such a lifestyle. The anxiety inherent in constantly grasping to obtain and maintain that way of life creates a state of misery. Instead, happiness can best be reached by setting our sights just on the things really and truly needed, which are few enough that they are not difficult to achieve.

Like most ancient philosophers of robe and beard, Epicurus articulated a view on death. It is perhaps his most famous piece of writing. He asserted that a proper understanding of life's end will help us reach a state of calm and happiness right now:

> Grow accustomed to the belief that death is nothing to us, since every good and evil lie in sensation. However, death is the deprivation of sensation. Therefore, correct understanding that death is nothing to us makes a mortal life enjoyable, not by adding an endless span of time but by taking away the longing for immortality. For there is nothing dreadful in life for the man who has truly comprehended that there is nothing terrible in not living. Therefore, foolish is the man who says that he fears death, not because it will cause pain when it arrives but because anticipation of it is painful. What is no trouble when it arrives is an idle worry in anticipation. Death, therefore—the most dreadful of evils—is nothing to us, since while we exist, death is not present, and whenever death is present, we do not exist. It is nothing either to the living or the dead, since it does not exist for the living, and the dead no longer are.

Epicurus was an atomist: He claimed that everything was composed of small particles (the atoms we know today weren't identified until the nineteenth century). We die, in Epicurus's view, because these atoms become disarranged; after death the atoms continue this path of decay—what physicists would now call entropic dissipation. Our mind, which is linked to the body, as Epicurus seems to have concluded and modern neuroscience daily confirms, takes the same path. The paths of body and mind are one path.

The personal self as a unified, internal experience disappears with the body's demise. I had first settled this to my personal satisfaction when I was sixteen during a solitary walk in cold autumn rain. I emerged a nonbeliever after a struggle between the views I had been taught during my Catholic religious upbringing and what I had learned about the universe, stars, atoms, and evolution from studying and probing the interconnected knowledge systems of science. My psychological being will really terminate when my biological life is over—this is easy to say but not always easy to fully accept. I won't even be able to take satisfaction in knowing I was right? I won't be able to witness, looking back from either heaven or hell, what unfolds in humanity's future? After death, true nothingness would not even be blackness because, as Epicurus realized, there would be no sensation and therefore death would be neither darkness nor light. It is the nothing of nothingness. What a contrast to the view of a real afterlife, as assumed in this joke:

> A man saw that his death had been reported in the newspaper, obviously a big mistake. He called up a friend and said, "Today's paper says I'm dead!" The friend said, "I saw that. Where are you calling from?"

Epicurus embraced the fact of nothingness as a foundation for happiness. Because there is nothing, neither pain nor pleasure, in death one should not be concerned about it. You are alive now, and that's what counts. When you are dead you won't be worrying. Problem solved! Don't worry now.

Epicurus, I submit, had the best answer to the question about how to approach the issue of inevitable mortality, and more than two thousand years ago. He still has the answer. Time, perhaps, to end this book!

When I recently attended the memorial service of a friend, many assumed the existence of an afterlife. These folks referred to our mutual friend as somewhere in another realm playing music or watching us speak about him, even laughing at us, and, because my friend was an accomplished philosopher, they envisioned him at that moment talking, yes, even arguing, with the great philosophers in heaven.

John Richards, a professor at West Virginia State College, was a hero of mine. How I hurt that he's gone. He died too young, at age 50. A heart attack dropped him at home. He never even made it to the phone.

Of course, I didn't try to counter the mind–body dualism some exhibited at the service. We fine people—and at least for the moment death had made us all fine in our common respect and humility—were bonded in shared mourning and affection. The theme was how John had improved our lives. Former students gave heart-wrenching testimonials about how John had enriched their lives, how generous he had been with time, how he famously teased them (a particular sign of endearment in Appalachia), and how he encouraged them to trust themselves. A dulcimer club had been reborn through his efforts. A well-known novelist recounted their wide-ranging conversations. A childhood friend said he had lost "the coolest man I've ever known."

The philosophy of Epicurus is recognized for its focus on friendship as much as its sober, edifying view on death. "Friendship dances around the world proclaiming to us all to rouse ourselves to give thanks." To the man for whom death was nothing, friendship was everything: "Of all the things that wisdom provides for living one's entire life in happiness, the greatest by far is the possession of friendship."

It can be empowering to recognize that death is nothing and friendship (or love) is everything. The couplet helps plant oneself firmly in the here and now. This is not a navel-gazing here and now but a presence that extends outward into the sacred dance of all life, to believers and doubters, to male and female, to old and young.

The sense of larger social connection we had at John's memorial service came of being there with particular people who possessed special personal histories. Through expression of individuality, we revealed a common humanity. In this way, memorial services involve not only the particular but the universal.

All of us share certain universals about death. They include, of course, the fact of our own personal mortality. But they also include other common properties, born from the roots of our psychological and cultural evolution and from the deepest taproots of our biological evolution.

Those common properties don't stop with ourselves, our species. In death as well as life, science has shown us, we are bound to all beings in a larger order. A broader recognition of death gives us an opportunity to see who we are on the universal stage and how intertwined all life is with death: To find where we came from in the nearly 4,000-million-year-old pageant of evolution. To see our lives as made from cells that live and die in our bodies. To ask why the human lifetime is what it is. To explore how the fear of death affects daily life, and how death helps to create, maintain, and push forward the magnificent course of life. If Epicurus is right that there will be nothing when we are gone, as I think he is, let's join in that dance of his, revel in our connection with all of reality, and strive to shape this life right now with creative intensity for all as best we can.

PART TWO

Evolving Life, Evolving Death

⟶⊙⟩ 3 ⟨⊙⟵

Origin of Life as Origin of Death

Astrophysicists don't even blink at their metaphors when they speak of the births and deaths of stars. A star's mass at "birth" determines its life span. Our sun, a middling-size star, in about 5,000 million years will enter a stage in which it first rapidly enlarges to a red giant and then collapses into a smoldering white dwarf. The sun as we know it will be history, though it will smolder like embers in a dead campfire for additional eons.

More massive stars end even more dramatically, like fireworks. When they explode into brilliant supernovae they scatter nuclear ashes and chemical elements that can coalesce into future stars and planets; they serve as the essential raw materials even for any alien living things somewhere "out there." All the carbon in our human bodies, for example, is nuclear ash from ancient stars that lived and died before our sun.

What about the known, visible universe itself? It was born in a Big Bang nearly 14,000 million years ago. If it continues to expand until all matter is cold and dispersed, as some scientists believe it will, the universe will finish in a "cold death." Astrophysicists have also dubbed this (ignoring some details of thermodynamics) a "heat death," because temperatures will be everywhere uniform and in a state of maximum entropy.

The universe, stars, and other large-scale physical systems such as galaxies, continents, or mountain ranges—all lend themselves to biological analogies of birth and death. These systems form, remain in states that are more or less stable for some periods of time, and then fade or break apart—thus ending their "lives." The long time scales involved for most of these large physical systems—millions to thousands of millions of years—provide a relatively fixed tableau for the more rapid-fire comings and goings of living things. The physical giants are virtually immortal compared with us.

We live sandwiched between the near-immortals of these very large systems and physical Methuselahs of another kind—those at the other end of the size scale. The metaphor of birth and death is not usually extended to the atoms, but we can easily do so. We could, for example, say that hydrogen atoms "die" in the cores of stars when hydrogen nuclei fuse into the slightly larger ones of helium. Further fusions, or deaths, in turn give birth to carbon, oxygen, and all other nuclei of larger elements. These are the atoms flung into space by supernovae explosions of massive stars.

A typology of the kinds of deaths of atoms would also include radioactive fission transformations of unstable atoms, either by natural decay or in nuclear reactors that generate electricity. When we survey all these types of atomic births and deaths, it is clear that, for the most part, atoms vastly outlive us.

Most atoms are, like the galaxies, near-immortals. The single protons at the centers of the hydrogen atoms in the H_2O you drink today were formed a mere few minutes after the Big Bang, when the cosmos cooled enough to allow the formation of stable small atomic nuclei. Thus the hydrogen atoms (their electrons can come and go, but the hydrogen is still hydrogen) have existed for nearly 14,000 million years. The one oxygen atom in that same H_2O molecule, like an atom of carbon in your body, was forged in a star that exploded earlier than the birth contractions of the sun and its solar system. Oxygen atoms have thus not been in existence for quite as long as most hydrogen atoms. But still, our oxygen atoms are older than the Earth's 4,500 million years and have participated for all that time in the roiling dynamics of Earth's geology and biogeochemistry.

The relative immortality of the very large and very small serve as a stable stage set for us. Compared with stars and atoms we are the briefest of candles, temporary patterns of peculiarly organized matter, blinking on and off upon the surface of a huge planet bathed with streams of steady high-intensity sunlight, with bodies composed of super-stable atoms that can be rearranged to create me, or you, then a soil bacterium, and then, say, a traditional symbol of ephemerality, the Japanese plum blossom in spring.

Our bodies themselves can be thought of as a set of imper-manent arrangements of permanent things. The biochemically active molecules inside our cells, though made from atoms, come and go with blinding frequency. The electron bonds between atoms in molecules are easily broken and re-formed. These bonds link atoms into temporary small families or large corporations. Many chemical reactions are quite lively at Earth's surface temperatures. But as cauldrons for creative control of the rapid-fire births and deaths of molecules, above all are the living bodies of organisms. To the molecules within them, creatures are like the relatively immortal universe itself.

This layering of the shorter-lived molecules within the longer-lived organisms has much to do with the source of the metaphor that we project onto the objects of the cosmos: biological death. Death occurs when the internal molecular dynamics cease. An organism is more like a whirlpool than a mountain.

With the advent of biological life, we'll eventually see, some aspects of death became a vital adaptation of living things, as vital as sexual reproduction is to mammals or butter-flies. But before we get to the Grim Reaper's more elegant evolutionary manifestations, we should approach him in his most stripped-down form, which came into being so long ago without any fancy design. Death just simply had to be when life itself arose.

The origins of life are shrouded in substantial mystery, but there are clues about its timing. All of life is water-based, and so life must have required water to begin. Sedimentary rocks show that water was present on Earth's surface by about 4,000 million years ago. Geologists' findings from the field can conjure general conditions of the early Earth: Occasional but massive bombardments from rogue objects in the solar system followed Earth's contraction from a cosmic gas cloud into its solid, round body. Either life began after this interval of massive bombardment, or it originated earlier and made it through the periods of impacts by living already dispersed throughout the ocean, in deep-sea vent communities, or, tantalizingly, deep in rock (a realm where even today we can find bacteria living). In the view of some scientists the evidence for early life, gleaned from carbon atoms still found in some of the earliest rocks, has pushed back the time of life's origin close to the final stage of the giant impacts. But it could have arisen earlier and we would have no geological record of it, yet. Much controversy and uncertainty remain because geologists are faced with extremely ancient, scant evidence. The ocean's deep volcanic vents, because they harbor certain kinds of natural chemical reactions that are found in the "core metabolism" of organisms, are one primary candidate site for the geochemical womb of all future creatures. But perhaps—and this is debated by serious scientists—microbial life came to Earth from elsewhere as inadvertent hitchhikers on interplanetary meteorites. Perhaps the spore-harboring objects were ejected from impacts on Mars, because that planet would have cooled to hospitable conditions earlier than did Earth.

Although the when and where of life's origins remain enigmatic, the origin of biological death is quite certain. It began with life. Simply put, life's very fecundity made it impossible for every microbe to live forever. Microbes—life-forms best seen with microscopes—are like us. The processes of growth and reproduction common to every dynamic living thing require capacities to draw regularly upon concentrated kinds of energy and to gobble up matter that can be converted into more of the stuff of the living self. Without death, this growth and reproduction would lead to an explosion of an infinitely increasing mass of life.

Bacteria are especially tiny microbes. Typically, tens of thousands could tile the period at the end of this sentence. In the earliest era after the origin of life, probably only simple cells at that microscopic scale existed, in the class of cells called prokaryotes. Prokaryotes are unlike the larger cells of our body and those of plants, animals, fungi, and many single-celled organisms. Prokaryotes lack a defined central nucleus and other internal, membrane-bound organelles. (Archaea, or archaebacteria, a genetically distinct ancient lineage of bacteria-size simple cells, are also prokaryotes, but for our purposes the distinction isn't critical, so with one exception in the text I refer to both by the simple term *bacteria*.)

Bacteria are everywhere and numerous. A microbiological census has pegged their numbers globally—including those in the oceans, the soils, and even inside plants, termites, birds, and other actors in the biosphere—at several hundred billion billion billion (10^{29}). On the skin of your human armpits and groin dwell about a million bacteria per square centimeter. Fortunately for those at all squeamish about so many little critters happily nestled in our crevices, most of the skin's surface harbors only several thousand bacteria per square centimeter. Mouths, gums, and colons, though, are more rife with them.

Bacteria are also fecund. Let's look at the sci-fi nightmare in which an initial single bacterium reproduces unchecked by death. In reproduction, a "mature" bacterium typically divides in two. The two offspring cells then each grow until they reach their mature sizes, about double their birth weights. These each in turn reproduce, essentially splitting into two. If we start with a single bacterium and if the doubling rate is once an hour, then simply by regular doublings the total mass would swell by geometric leaps into the mass of a typical human in 2 or 3 days. And some types of bacteria can reproduce every 20 minutes in laboratory conditions.

The cycle of growth and division for many types of bacteria in wild soil or natural water may proceed far more slowly than the doubling time just suggested. One global estimate for the average doubling time is on the order of once a year. A single bacterium could then grow into a human-size colony in 60 years. So what's the problem? Well, in the absence of death, it would only take another 60 years or so for the bacterial mass to snowball—or cosmically avalanche—into a mass equal to the entire Earth. Even if the whole planet were edible, at the far-fetched point at which bacteria completely consumed the Earth, the colossal colony would then die. But most of the planet is chokingly inedible, so bacteria would die of starvation well before this ultimate limit in resource depletion. The same story holds for local sites of limited resources within the biosphere—bacteria must die, at the very least because they run out of resources on which to feed, or they succumb to any number of other adverse environmental conditions.

Sometimes, instead, hardships induce bacteria to beat a metabolic retreat into dormancy, into tough time capsules called spores. Not all individuals in any single colony will make it into spores, and many types of bacteria are genetically incapable of spore formation. But once in a spore, the lucky bacterium can remain viable for ages. One famous rejuvenator, a Rip Van Winkle bacterium, was two hundred thousand years old when resuscitated from ice in Antarctica. More controversial oldsters are reputed to approach two million years in a dormant state. They are not dead, just barely existing, their usual internal molecular whirlpools nearly frozen, their animation suspended. It's a model we might like to emulate for ourselves, eventually, with biotechnology or other tricks. Science-fiction writers and futurists, for example, have often postulated the potential to deep-freeze a body and then return it to life in the future, say when a particular disease has been cured. Perhaps, in journeys that might take thousands of years, we will desire to travel to the stars.

The world cannot fill to the brim with just spores, either. Something very basic must happen to most of the bacteria that actively live or even go into spores. They must not only die. They must disappear.

In today's world, bacteria have many predators. In dying, others live. In dying, their bodies transform into potential sources of nutritious matter and free energy for other creatures. For example, larger and more complex single-celled protists in waters and soils consume bacteria. And underground, the thin, white hyphal threads of fungi release digestive enzymes as chemical weapons into cities of bacteria and then sop up the dissolved remains.

Some bacteria are preyed on by other species of bacteria. Predatory bacteria can secrete toxins to attack prey bacteria. Like the fungi, these predatory prokaryotes digest their prey externally, then transport the nutritious liquefied remains into their bodies to support their own growth and reproduction. Some predator bacteria are very small compared with the bacterial prey. They attack their way into the bodies of larger prey and reproduce exponentially within. Soon the hapless hosts burst open and spread great numbers of the next generation of predators.

Viruses, too, should be considered predators. Most use a similar sequence of entry, takeover, and internal replication, ending in a final bursting of their host cells. This means of replication by killing is how specialist viruses called bacteriophages (bacteria eaters) make their "living." Lacking any internal whirlpool of metabolism, viruses are essentially protein-protected bits of DNA or RNA, with only a tiny number of genes. They are not able to replicate by themselves in a bath of just chemical nutrients. Instead, they force their way into bacteria (or today, into cells of any creature), then use their host cell's own inner biochemistry to churn out multitudes of more viruses. Essentially they hijack a manufacturing plant and reprogram the machines to create more hijackers. In today's oceans, viruses are recognized as significant killers of the prokaryotes, thus limiting bacterial populations. This was probably true near the origin of life as well.

What we see in this tale is little evidence of modern morality. The living survive and thrive on the death of others.

At base, it's a huge power grab. Now, to an electrical engineer or physicist, power is defined as an energy flux over time, and organisms do require energy in flux, to resurrect their proteins that are constantly degrading in the thermal sea of entropic reality. I've at times thought, when quietly taking in the scenery at, say, the side of a small mountain stream, how many of the creatures around me, from vultures to mice to soil bacteria, would just love to feast on me. Were I to fall, they would not cry a single tear for me. Indeed their mouths, those that have mouths, would salivate at the sight or smell of my body. And the pores of the bacteria nearby would be poised to exude digestive enzymes at first touch of my corpse, whether still warm, or cold and grossly odoriferous.

Recycling of the Dead

U nless you are buried in layered shells of hermetic boxes and embalmed like the pharaohs of ancient Egypt, bacteria will get you in the grave and transform the carbon in your body to carbon dioxide, the main molecule from which you were built in the first place. That is because plants eat the air's CO_2 to create the array of carbon-based organic molecules that feed animals, including you. Whether you are a raw-foodist, vegan, lacto-vegetarian, omnivore, or mostly carnivore, like the Arctic-dwelling Inuit, you thus depend, ultimately, on atmospheric CO_2.

Should you choose, you can bypass the bacteria by cremation. This also returns your carbon to atmospheric CO_2. So whatever route your body takes after death, you end up as a molecule that is the primary essential nutrient for all photosynthetic algae and plants on Earth. So if you are ever feeling low about not having a purpose in the immensely humbling note of your impermanence . . .

You also are a source of carbon dioxide all during life. In order to survive, you must always be converting some plant or animal life into that waste gas. You might eat living cells of lettuce or apple, or cells of other living things killed before or during cooking. You process the food, derive both matter and energy from it, and produce CO_2 as a waste gas that leaves the body with each exhalation. Carbon also exits in your liquid and solid wastes, which then are turned into CO_2 by bacteria in soil or solid waste processing facilities (your local one is worth touring, I assure). You are a CO_2-production factory. By creating death you live.

A carbon budget shows that every six months you exhale an amount of carbon (in CO_2) equal to the weight of your body's carbon, totaled up from toe to head. But a far greater amount of carbon is turned into CO_2 so you can feed and exhale. Growing food crops produces waste carbon in the parts of the crops that die after we harvest the grains or fruits of their photosynthetic labors. And every gram of carbon in animal protein requires many times the equivalent in plant mass (or even other animals) digested by that animal destined for our tables. The average Western diet requires the deaths of plants about equal to your bodily carbon each month.

Carbon is an element in one of the most crucial of Earth's "biogeochemical" cycles. Also vital are the cycles of nitrogen, phosphorus, potassium, iron, and all other elements essential to organisms. Each element has its unique story. Consider, for example, the essential component of all proteins: nitrogen. We can begin one version of its labyrinthine saga with nitrogen gas that passes from atmosphere into the soil. There the gas is consumed by nitrogen-fixing bacteria that excrete the nitrogen atoms in ammonium ions, which are perfect for the nutritional needs of plants. Plant protein is then consumed by animals that convert it into their specific cell proteins. We, as mammals, are obliged to periodically excrete nitrogenous wastes, such as the uric acid in urine. Back in the soil (or waste treatment facilities), the discarded waste forms of nitrogen are converted by still other forms of bacteria into ammonium ions again and also into other kinds of nitrogen molecules. Still another type of bacterium thrives in soil or ocean sites with very low levels of available oxygen. These microbes convert the nitrogen ions back into nitrogen gas, which is their specific waste, thus completing the grand cycle.

All the various biogeochemical cycles contain transformative steps that convert one form of a chemical element into a different form. Many of these steps involve the deaths of cells, of parts of living things, or of entire creatures.

But not all wastes and deaths are recycled. Some are buried. When the geologists talk about "carbon burial" or "nitrogen burial" they don't mean human bodies in graveyards. Any dead creatures within soil are still within the larger system of the biosphere. In soil or air, a carbon atom, to focus on that element, is still biospherically "alive." As carbon passes through stages that convert it into various forms—in air, soil, water, and living things—it remains within the biosphere system. Its typical lifetime in this special system is about 100,000 years. The biosphere is so tightly closed that a carbon atom can be kept circulating within the confines of Earth's surface fluids for a very long time indeed.

When carbon ends its "lifetime" in the biosphere, it doesn't stop being carbon. It merely passes into a deeper zone. One is reminded here of ancient myths that feature souls, victims, or heroes descending into the underworld, as a dramatic moment in the story. Like those mythic souls presumed to continue to live but in a new form, so carbon transported downward and outside the vibrant biosphere, after "burial," continues to be carbon but somewhere deep and dark, and often hot.

Carbon is buried as detritus from dead marine plankton when it fluxes out of the dynamic surface system in the form of tiny calcium carbonate shells. The coal we mine to burn for electricity is the dead and highly compressed remains of giant ferns and mosses from dinosaur-era swamps. Our precious, diminishing reserves of oil were long, long ago the sediments underneath some of the world's most productive marine areas ever. Verdant patches of algae grew, then fell into the sediments at such rapid rates of death that even the voracious bacteria alive there could not keep up with the rain from what was their heaven. The sites and rates of death that led to the fossil fuels upon which modern civilization came to depend were historically unique burial traps.

More commonly, carbon that was buried from organic tissue in the form of the bodies of plankton was finely dispersed. Today we see it as the black tincture in rocks such as shale, in contrast with the pervasive white of limestone rock that entombs once-living carbon in a paler shade.

All these buried forms of carbon can eventually spring back up, like the ancient Greek myth of Persephone emerging from the underworld to bestow life to the surface. She was said to rise up annually, as a rite of spring. But carbon's stay below is typically millions to hundreds of millions of years. Its ports of reentry are the volcanoes and surfaces of rocks that dissolve when exposed to soil, rain, and weather, thereby returning carbon to the surface circulation of active cycles.

How dependent is life in the sunny biosphere upon this resurrected carbon? In the long run, very dependent. Without the reemergence—a kind of biogeochemical reincarnation, if you like—all carbon would slowly and surely exit from the interconnected surface system of life, air, soil, and water. Emergence would be limited to only truly primordial carbon that comes up as a portion of volcanic activity.

The flow of all reincarnated carbon is not nearly enough to supply the hungry stomata of plants and the ravenous membrane pores of algae and other phytoplankton—all of which require CO_2. For perspective, it's worth comparing the annual need with the supply.

The total amount of carbon taken up into the bodies of photosynthesizers—living things that perform photosynthesis, the conversion of carbon in CO_2 into carbon-based organic molecules at the heart of all life—is 100,000 million tons every year. That amount is roughly split forty–sixty by basically all marine phytoplankton and terrestrial plants. For contrast, the amount that comes up from the dark depths below is merely

500 million tons every year. The annual photosynthetic need dwarfs the annual supply from below. The differential is a factor of 200. Plants and phytoplankton require carbon at a rate 200 times greater than provided by the geological underworld.

Where do they get it? The simple answer is from water and air. But where do water and air get carbon? It comes from you and me, from all other mammals, from fish and frogs, from caterpillars and butterflies, from worms, and most of all from bacteria in the world's soils and oceans. It comes from the CO_2 these metabolically equivalent "respirers" expel and exhale.

The cycle that requires that living things cause death is the means by which life-forms get what they need. The factor of 200 we derived above is the measure of amplification given to global photosynthesis, and thus, as carbon goes around, to all other organisms, by the ongoing cycle of death–life–death–life, because without the cycle photosynthesis would be limited to what it could draw from the tiny supply of CO_2 from the geological depths.

A hefty 99.5 percent, therefore, of the carbon as CO_2 taken in by plants and algae comes from creatures that either cause death or feed upon the already dead and produce CO_2 as waste. Without recycling the dead, the annual growth of phytoplankton and plants would be limited to the meager gift of carbon flow coming up from the deep underworld. Imagine a world in which the forests were a mere two-hundredth of what they are today, or the oceans only a two-hundredth as productive.

My personal enthusiasm for these biogeochemical cycles comes not just from the fact that they offer a fascinating opportunity for scientific investigation. They also allow me to embrace a personal form of immortality. I become other creatures through my atoms. This doesn't just happen through those creatures at the streamside who await my death. Even more so, I contribute to life's biogeochemical cycles with my every breath, with my conversion of the cells I eat into the CO_2

I exhale that returns and helps fuel the next rounds of photosynthesis (and thus me).

But—is this all? Whatever your reaction, we have gone far beyond Epicurean logic that "death is nothing to us." The ancient philosopher was focused on psychological anxiety about mortality. But consider how life's very exuberance— the prolific growth of trees, grasses, algae, and thus salmon, bears, and eagles—is dependent on the death–life cycles that reuse carbon and other elements. It's a world of biogeochemical "friends." We witness a pattern of death that is more than neutral nothingness. The pattern might itself be a thing to be celebrated. In any case, we live thereby.

Suicidal Bacteria

The noted astronomer Carl Sagan, who penned words used as an epigram at the start of this book, described the power of death in the process of evolution: The "secrets of evolution are death and time—the deaths of enormous numbers of life-forms that were imperfectly adapted to the environment; and time for a long succession of small mutations that were by accident adaptive, time for the slow accumulation of patterns of favorable mutations."

Sagan brings death front and center to evolutionary creation. No death, no evolution. No evolution, no production of new species from the cauldrons of material selection that led to us and all the other past and present marvels of the living biosphere.

Death and life turn the global biogeochemical cycles that enhance the total number of active living things. Death and life create evolution and thus the constant births of not only new generations but also new types of creatures. Those new types have new characteristics, one of which can be new forms of death itself.

This is amazing: From the creative cauldrons of selection, which require death, have come forth, even in the simplest cells of the prokaryotes, adaptations that incorporate death as a part of the very functional aspect of survival. At first blush, this seems counter to what would be required for evolutionary success. Death as a functional design? How could this work?

To be clear: I don't mean the kinds of evitable deaths that

occur from resource limitations or from the webs of predators and prey that create the wonders of the biogeochemical cycles of carbon, nitrogen, and other elements. These deaths, though necessary in the course of evolution, are by-products in the sense that they are not a part of the actual selected physical or metabolic adaptations of the creatures that die. What I'm getting at instead are cases in which death becomes part of a creature's evolutionarily selected adaptations of survival. Some examples should clarify.

In bacteria such as today's genus of *Myxococcus,* also called slime bacteria, we see that somewhere along the paths of the multiple lineages spun forth by evolution, death took a great leap forward. Death took on a new kind of life, because it became so integral to life.

An individual cell of the species of *Myxococcus xanthus* looks like a tiny cut-off finger. It's a rod-shaped bacterium found throughout soils and especially in the detritus of dead, decaying leaves. The rods have the ability to hook together end-to-end and form colonies that can glide as coherent groups: They've been called the bacterial equivalent of a wolf pack.

Most remarkable for our discussion is what happens when the myxococcus bacteria become environmentally stressed, say from lack of nutrients. They gather into an even denser colony. Signals among the members of the swarm—which may number a hundred thousand or even up to a million individuals—induce them to well up into a pyramidal mound. As the bacteria rise up they differentiate into two major types. Some become a stalk; the others form themselves into a ball and will develop into numerous tiny spores on top of the stalk.

Think of the water towers seen in many American small towns. Each tower typically has a tube that comes up from the ground and rises to a giant bulbous container that the tube supports. The bulb stores water, and its height provides gravitational pressure as the water flows downward as needed

by the townsfolk. In the developed, multicellular myxococcus colony, the ball-on-stalk structure is tens of thousands of
times smaller than the water towers but visible to the eye at 1
to 2 millimeters tall. (In some species, the support stalk more
closely resembles a tree trunk with multiple branches and
"fruits" rather than a single column topped by a sole "bulb.")

Cells in the spherical bulbs or fruits form myxospores,
while cells in the supporting stalks will die. Death occurs by a
process known as autolysis. Autolysis is self-dissolution, and in
the case of myxococcus the process is required for the formation of the stalks. The function of autolysis is not yet known,
but it might be a way of supplying additional nutrients to the
cells that will become the spores. The stalk definitely provides
support.

Here, in tiny bacteria, death has actually evolved into being
an adaptation in the service of life. The supporting cells in the
stalks of myxococcus help the cells in the fruiting body to get
up off the ground. The millimeter or two of height enables the
spores, when released, to ride air currents that are surprisingly
stronger than those right at the surface. This distance, though
infinitesimal to us, gives dispersal power to the myxococcus.
The spores get carried to new and, in some cases, favorable
locations for new colonies on the dark and dusky forest floor.

This sacrifice through suicide works because the cells of the
colonies are basically clones (though the details of variants,
and the existence of genetically distinct individuals that wran-
gle their way preferentially into the fruiting bodies and avoid
ending up in the stalk, thus shirking self-sacrifice, are still under
investigation). When clones sacrifice themselves for the good
of others that are genetically, in essence, themselves, sacrifice
works: Death leads to reproductive success of their genes. The
little bugs are hard to kill because they've learned to die.

Cell biologists call this pattern of sacrifice "programmed
cell death." It is known to have been invented multiple times

in the evolution of bacteria. One other prominent poster child (or "poster bacterium") example is the species *Bacillus subtilis*.

In the bacillus bacterium, programmed cell death occurs during its unique process of forming a spore. The genome in the cell is duplicated and then the cell grows a wall that separates the two genomes into a larger, so-called mother compartment and a smaller compartment called the forespore. The mother cell cannibalizes some of its proteins to make special ones for the forespore and performs other acts of biochemical generosity before it disintegrates and dies. The complete forespore can now make it through unfavorable environmental conditions to germinate at some future time.

Death evolved in bacillus as part of a division of labor. There is no genetic conflict because both mother cell and forespore are clones. By dying, the mother cell is, in essence, helping its same genetic self in the form of the forespore.

Another example of useful programmed death in a microbe is yet another soil bacterium called *Streptomycetes* (the genus within which one species is the source of the antibiotic streptomycin). The reproductive sequence of streptomycetes is so complex that this type of bacteria is truly multicellular. Microbiologists have discovered that two separate waves of programmed cell deaths occur when the bacteria colony enters the phase of spore development. The end result is that cells that are slightly deeper die in support of the cells closer to the soil's surface. Their deaths provide both a physical support network and a supply of nutrients, all in the assistance of those cells that enter into the formation of spores. Spores will spread, increasing the possibility of landing genes in new and better environments.

Cell suicide has thus evolved multiple times as a stable and even powerful solution to the challenges of staying alive and propagating. In all these cases, the theme is division of labor in support of spore formation. What is crucial for the evolutionary story of death is that death is programmed as a primal differentiation of function. In these cases life evolved a new kind of death.

I say *new* form of death, because in these types of bacteria, which currently so fascinate microbiologists, certain cells senesce and die in an orderly way. Once there existed programmed cell death, then death itself in evolution became a feature of development that could be selected for or against, that could be honed and tuned up or down to fit the circumstantial niches of particular forms of life. As will become clear, this step was essential in making us humans who we are.

In recent years, furthermore, evidence has accumulated that an orderly cell suicide takes place even in solitary bacteria that apparently don't live in functionally differentiated colonies like myxococcus or streptomycetes. Free-living tiny plankton such as the photosynthetic cyanobacteria that fill the surfaces of oceans have been found to commit suicide when stressed by nutrient deprivation. In fact, many researchers are now starting to think that all cells have the genetic capacity to go into a suicide mode. (How much the genetic program is common to all cells and perhaps evolutionarily related, and how much was evolved convergently, is hotly debated and under intense scrutiny in labs right now. Its adaptive consequences are also under investigation.)

Programmed death was a significant step in the evolutionary history of these examples. With the advent of programmed cell death, death went beyond its position in evolution as a by-product of predation and cog in the biogeochemical cycles. Death had now become one of the genetically sculpted features of organisms themselves. Death and life had become one in a truly profound way.

6

Little Deaths in Big Bodies

In an oft-cited poem by Robert Frost, a traveler recalls two roads that diverged in a yellow wood. The scene served as a metaphor for a decision his younger self took to set a life trajectory. An image of diverging paths also evokes the way that evolution branches into different species or, given enough time, into whole different genera.

But sometimes the roads of evolution have merged as well. The process, a coming together of organisms of different species to form new organs, organisms, or species, is called symbiogenesis. Lynn Margulis, the mother of Dorion Sagan, author of the *Sex* part of this book, has tried to ensure that this process of merger has gained the scientific and public attention it deserves.

One such merger that took place in the prokaryotic realm of the very small became over time a major evolutionary event. It led to a fundamentally new kind of more complex single-celled organism and in so doing made possible the evolutionary development of much larger life-forms such as fungi, animals, and plants. This event also contained the seed of a new kind of programmed cell death.

In the event, according to Margulis, a first symbiogenesis of prokaryotes led to the earliest evolution of protists, cells with nuclei and new kinds of internal dynamics that were capable of ingesting other cells and therefore acquiring "genomes at a swallow." These ancient protists, coping with the rise of atmospheric oxygen, incorporated other prokaryotes. A second set of symbiogenesis events resulted in many descendants, including eventually us, whose cells harbor oxygen-respiring organelles called mitochondria. This big step, the origin of complex (known as eukaryotic) cells, preceded the evolution of all the life on Earth that we can see with the unaided eye.

The numerous mitochondria in eukaryotic cells function as sites that oxidize organic carbon compounds with oxygen drawn from the air or water. As a result, the mitochondria are able to produce high-energy molecules. These are used by the cells to drive the reactions that assemble proteins, such as the enzymes and structural molecules. Such so-called mitochondrial power plants are essential to the lives of most complex cells, from ciliates such as paramecia to all cells of fungi, plants, and animals, including us.

The mitochondrion also has what might at first seem to be a dark side. The tiny organelle plays a death role so profound that French scientist Guido Kramer has called it the "central executioner."

This biochemical executioner has been most studied in the cells of vertebrates. The membranes of their mitochondria become permeable at an early stage of the programmed death. Specifically, a protein called cytochrome-c leaks from them. This is the point of no return in the cells' suicide programs and exhibits an unusual example of biochemical irony. Cytochrome-c is also essential for minute-to-minute cell life. Inside the mitochondrion, cytochrome-c transports an electron between steps in the "power plant" cycle that forges the high-energy molecules. And yet cytochrome-c is crucial in the path to planned cell suicide. The leak does not itself cause the death; rather it initiates a series of signals that result in self-execution.

It now appears that mitochondria play a role in some of the programmed cell death machinery in plant cells as well. But why would any of the complex eukaryotic cells employ a suicide program? Perhaps parts of this eukaryotic program were inherited from prokaryote death programs, such as the single-cell cyanobacteria noted at the end of the last chapter. There are tantalizing clues of some true ancestry of parts of the program in deep time. But ancestry is one thing. Actively functioning, complex biochemical links and wheels will not be maintained and used by later evolutionary forms unless those later forms also require the links and wheels, perhaps highly modified, to live.

One clue to the functional importance of programmed death in eukaryotic cells is the fact that there are differences in how these cells commit suicide. Some use the ancestral pathway of death granted by the mitochondria. Other cells have evolved different tricks to ensure death at the proper time. These differences, to me, are particularly revealing. They show that death as a functional process in evolution has itself been subject to honing by evolution. Death was so potent a force in creating viable forms of life that its invention was accomplished many times in different ways. Even within the human body, as will be discussed, some cell deaths are driven by the mitochondria as central executioners, and some cell deaths are accomplished by other means.

In what is probably the most celebrated example of controlled cell death during human development in the womb, the hand of the fetus changes from a flat paddle to separate digits. Deaths that involve the executing power of mitochondria eliminate cells in the in-between zones of what will become the future fingers. Cell death is here a sculptor of development. Unlike a sculpture of marble or wood, in which the deleted scraps lie around the floor to be swept up and thrown away, in this cell death process the material is recycled into other cells.

A sculpting by controlled death is also necessary during the development of complex bodies such as frogs. Coordinated chemical signals among cells can be commands to die. One classic example occurs as a tadpole turns into a frog. During metamorphosis, the tadpole's thyroid gland secretes a hormone that induces the cells of the tail to die. The tail is eventually reabsorbed and so recycled into the body of what becomes the four-legged, tail-less little frog.

Cell death during development controls not only gross morphology but also the numbers of cells in many tissues of animals. Cells are typically overproduced and then culled down to the necessary final numbers. The best-known example of this occurs in the development of brains.

We originally grow about twice as many fetal neurons as eventually survive in our functioning brain. When these neurons develop connective networks, they compete to reach particular target cells. During the process, the target cells secrete survival chemicals for the neurons groping toward them. Those neurons that reach the targets live; they have received enough survival juice. Those that fail to arrive die. So in this case, the cells die because they did not receive the signals enabling them to live. This method of assembly serves as a quality control, both on the final number of neurons and as a way to eliminate those that aren't well connected.

The human body is replete with death in the support of larger life.

First, consider a number: In the adult human body, one hundred thousand cells die every second. This also means that approximately one hundred thousand cells are newborn each second. From these numbers we can deduce that on average the body's cells turn over about once a year. The turnover numbers show that life is a renewing pattern of matter. The atoms can completely change over, but our bodies sustain their identity.

The rates of this cell turnover vary crucially for different types of body cells. For example, the brain's cells turn over hardly at all. This stability is probably linked to the feeling of continuity we have as individuals, so intricately connected are the cells to one another in vast nerve networks. When nerves die, their connections to other nerves in the network vanish, with no guarantee that replacement cells can make the same complex connections. Perhaps memories would shift were too many brain cells often replaced with ones that made different patterns of connections!

In contrast with the brain, other places in our bodies are hot spots for rapid cell death and birth. Our skin is one. Its renewal is most apparent in the healing of cuts and scrapes. But the renewal goes on continuously. Everywhere in the skin, below the outermost, scaly layer, new cells are produced by division. The body has a neat trick here to ensure death. During the development of the outermost skin cells from the underlying living layer, the cells are enucleated (the nucleus is removed). Without central DNA, they are unable to produce fresh internal enzymes as old enzymes degenerate. They are doomed to short lifetimes.

The dying skin cells grow coats of an especially tough protein called keratin and lock into geometric positions alongside their neighbors whose borders fit together like dovetail joints. At the same time these cells are pushed outward toward the body's surface, where they eventually become the skin we see and feel. There they undergo the ravages of the environment. They are sacrifices for life. They are eventually marched right off the body in continuous shedding. But before being shed they do serve as an exquisite protective envelope. A layer of functional cell corpses coats our body.

The gut lining is another hotbed of frenetic cell renewal, with a turnover rate of less than a week. It's a dangerous place for a cell. Food is digested with an acid bath of powerful, molecule-ripping enzymes. And there are legions of bacteria there, some of which may be pathogenic. The gut cells are living right on the edge, using chemicals to tear apart things that are not that unlike themselves. The place reminds me, as a social analogy, of a coal mine. Workers in the extractive industries often operate at the edge of habitability. They work in dangerous conditions, to bring raw materials into society. They contract occupational diseases, and are injured, sometimes fatally. Laws have been increasingly more protective of the workers' rights, but in the early days (think of a Babylonian copper mine) workers were, well, expendable sacrifices, like the gut cells.

Then there are our beautiful red blood cells (somewhat like doughnuts, with middle "hole" zones just thinned, not completely punched through). They survive for only about four months after their production by division and, like skin cells, they don't have nuclei. They are not going to reproduce, and their days are numbered. Their lack of nuclei curtails their protein-making capabilities, and consequently their lives. When the red blood cells die of old age, in essence senescing as they lose functions they cannot maintain without nuclear DNA, white blood cells in the bloodstream find these dead husks and ingest them. Thus we recycle parts of ourselves.

The immune system also operates via a continuous stream of cell death and birth. The ravages from serving as internal defenses for the body require this. Those cells of the immune system in our blood that ingest and kill viruses and harmful bacteria have only limited lifetimes themselves. The immune cells are utilized—or remain unused too long—and then they die; in this case the central executioners, their mitochondria, guide these suicides.

The outer cell membranes of some immune cells are coated with specific types of molecules that can recognize certain enemies. Hordes of these cells fill ranks for some specific fight, say, against the flu. Other types of immune cells are born with their recognition molecules varied, which provides a way to search for harmful microbes whose features are not yet known but, in a sense, are guessed at by the immune system. In addition, some immune cells born from this speculative shotgun approach of random variation are potentially dangerous, because they code for attacks on the body itself, like a missile suddenly turning back on its launch site. These, too, are normally eliminated; but autoimmune diseases can result when the deaths of the vagrant cells are not carried out swiftly or neatly enough.

Body cells have different life spans that reflect their lifestyles. Nerve cells generally are long-lived. So too are the muscle cells of the heart. Pathologist Edmund LeGrand suggests that the reason for this is similar to that for the longevity of nerve cells. The critical timing of the heart, which requires its myocardial cells to conduct electrical impulses and contract, and all within a highly coordinated system, provides little tolerance for replacements that could be ill fitted for the network. It would be like suddenly shoving me into replacing guitarist Eric Clapton, with thirty seconds to go until showtime at Madison Square Garden. Better that Clapton himself has a long life and continues to rock-and-roll.

The cells of the heart and brain need to be protected, then, because both kinds operate in highly coordinated networks with other cells of their respective organs. But the skin cell on my hand isn't directly related to the skin cell on my cheek. The red blood cell now traveling in my right shoulder doesn't relate to the red blood cell in my left little toe, at least not in the same sense that the interconnected heart cells and brain cells coordinate with others of their kind. The greater or lesser requirement for protecting the integrity of every detail within a subsystem in the body seems to mandate which cells have longer or shorter lives.

All this death in the support of life is impressive in a mammal body such as ours. If we are a whirlpool through which matter and energy flow in and out, within that human whirlpool much matter is recycled. In this world of the body, death is a necessity in the development, growth, and even stability of life's whirlpool.

7

Built from Death

Surpassing in some ways the wonders of death within the living animal body are the roles of the functional dead in sculpting the towering lives of trees. If you go inward from the bark, past a thin layer of cells called the phloem and another narrow layer of cells that are actively reproducing, you come to a notable layer called the xylem.

The xylem consists of tubular columns of dead cells that function to move mineral-laden water gathered by the roots up to the needles or leaves. Its special, dead cells are called tracheids. Tracheids (or, when grouped into units, tracheid elements) not only provide water and mineral circulation but also support the entire tree in its climb upward against gravity. Without tracheids there would be no forests or grasslands, no green life on land, except for some ground-hugging tiny mosses and a paltry soil coat of photosynthetic bacteria and algae. For not only do trees contain tracheids, so do all nonwoody herbaceous plants. Tracheids are in all stems, branches, and trunks of trees, in the shoots of grasses, in flower stalks (usually in their centers), and even in the veins of leaves. In all these instances, the dead are part of the living.

Without the evolutionary invention of tracheids just inside 400 million years ago, the land today would be virtually deserted. For more than 90 percent of Earth history, neither land plants nor their vital tracheids existed. And because tracheids are dead, in them we have an ideal example of how nature turns death into life to create organisms from cells.

Tracheids are functional cell corpses. In the development of tracheid tubes the living cells destined to become tracheids are emptied of their cell contents. They are killed, in essence. The empty cells, stiffened with extra cell walls arrayed with lattice-work that would turn our structural engineers green with envy, are then linked end-to-end in long, vertical bundles. Their ends dissolve and they connect into channels for the plant's vital transport of water and mineral nutrients. How do cells know when and how to die? How do their remains then take their places as operational tracheids? Do the cells perform the trick by themselves, in suicides? Or do other cells assassinate them?

The process is complex. The favorite cells to study for tracheid formation are in an annual flowering plant, the zinnia. Zinnia cells can be induced to develop into tracheid cells by plant hormones and other biochemical substances. The main events are these:

As the transformation begins, just inside the surface of its outer wall, a cell destined to become a tracheid lays down an additional shell of cellulose plus a glue-like molecule called lignin. This so-called secondary wall boasts exquisite patterns of rings, spirals, quilted reticulations, or other tracery that evolved for long-lasting structural stiffening. A few hours after the secondary wall begins to form, the membrane that surrounds the central, fluid-filled vacuole space—an internal anatomical feature of many plant cells—disintegrates.

Membrane disintegration is a critical point in tracheid formation. Enzymes that had collected in the vacuole's fluid, looking somewhat like a water balloon within the cell, are released into the cell's surrounding internal living matrix. These protein-dissolving enzymes attack specifically targeted proteins, while other enzymes disrupt the DNA and RNA. After these events, the dying cell is elevated in status in the tree or small plant to the fully functional rank of tracheid in the water-conducting channels of the xylem. The dissolved mush of cell contents is evacuated and recycled into other cells. Tracheid formation is the purest form of sacrifice in which the victim, the cell, becomes a permanent, essential transport organ for the whole living creature, the plant.

The xylem consisting of tracheids is not the only part of the tree in which death plays a starring role. A layer sandwiched in between the bark and the narrow layer of cells that are actively reproducing is just as important as the xylem: the phloem. Unlike the tracheids, the phloem cells are alive. But they carry death warrants: Most have been enucleated. Plants, in their evolution, discovered essentially the same technique in the control of death that animals did. We, for example, as we've seen, have enucleated cells in our skin and in our red blood cells, which also evolved to have finite life spans in support of functions within the greater whole.

The plants' phloem cells, without DNA and the attendant ability to synthesize new proteins, need help from neighboring cells in order to live. In trees, phloem cells age, die, and are replaced, often annually. While alive they perform the critical function of transporting viscous, sugary, photosynthesized food made by leaves into the stems, the trunks, and even down to the roots, which all need food to live and grow. But watch this: Upon death the phloem cells become part of the bark. They take on a second, critical job in retirement. A tree that lacks bark is a tree with a death warrant.

In the case of a tree such as a ponderosa pine, the bark is thick and, as the tree ages, is shed in giant, red-brown flakes. These flakes contain ultrathin layers, which can be peeled off and look like jigsaw puzzle pieces. The thin sheaths are the annual layers of formerly active phloem. The sheaths split as the tree trunk expands over the years, creating deep crevasses. Looking into them is akin to viewing the Grand Canyon from an airplane, where the annual layers of former phloem are the sedimentary layers of rock in the canyon. Trees have various ways of splitting the dead layers of former phloem to enlarge the bark as it grows thicker, and expert foresters can tell the species of trees in winter just by observation of specific patterns by which death sculpts the bark.

And so we can say that death makes life in two key places in a tree: in the bark made from dead phloem cells, and in the transport xylem that, when aged, turns into the structural wood. Death is useful; it is in fact creative, as sacrificed cells become active participants in a larger living whole, thereby helping all the tree's living cells.

Controlled cell death functions in other parts of plants as well. It occurs in thin cross-sectional zones where the stems of ripe fruits and dead leaves will cleave away from trees, in a manner that preserves the sealed integrity of the whole plant. Cell death is vital to many stages of plant development. In flowering plants, once the ovum (egg) has been fertilized, the petals, no longer needed, die. Anthers, as well, die after the pollen they grow and then store is released to the wind or to insect pollinators. As plant parts are used and then die off, some internal nutrient-rich portions of the cells are returned to the main body, a form of auto-cannibalism, a lesson in recycling that we could learn from in "killing" the artifacts we buy, use, and use up.

Annual plants, in their brief lives, die after reproduction. Sometimes the dying is controlled in such a manner that the seeds receive final boosts of nutrition from the senescing tissues. This sacrifice is of the whole for the part. But this part, the seed, contains the germ of another whole, because an entire plant can grow from it, an awesome phenomenon.

So plants are gurus in the use of functional cell death. Trees might be the most towering of all these green masters. The tree is one individual whole living being that incorporates both living and dead cells into a larger life. The living tree, in fact, is a being flush with useful death.

8

Extreme Senescence

If a modern snorkeler could take a time machine back to the ocean of 500 million years ago, he or she could paddle around and be almost as delighted as today, with the ancient animals and reefs, though bizarre (no fish, for instance). A modern hiker, similarly, could feel at home—though, again, the species would be strange—walking in a sylvan green canopy of 300 million years ago. But let them emerge in time a few hundred million years earlier to ancient ocean or land and then our time travelers would be severely disappointed. All the "recent" large organisms, on the evolutionary scale of time, could evolve only given the ability to control cell death, enabling their complex developmental patterns.

With large organisms a new face of death shows itself: the death of the entire multicellular organism through a species-specific average life span. Like the microbes thousands of millions of years earlier just after the origin of life, large organisms could quickly consume the entire Earth, were the Earth edible and the organisms able to reproduce unchecked. Death must exist. And so it does, often violently in the food webs of animals. Yet there is also a unique and important additional pattern to consider in the senescent degradation of the adult animal body.

Single-celled yeast do senesce. A parent yeast cell that produces clonal offspring cells by budding eventually loses its capacity to create new and viable buds. Thus a yeast cell has a kind of life span. Furthermore, the genes that can lengthen this life span are being investigated for clues about the cellular dynamics of our own rate of aging. But are there pieces to the explanation for the various styles and rates of senescence that go beyond the genes? How does a creature's ecological niche come into play in producing, by evolution, its species-specific aging with its progressive susceptibility to disease and other metabolic failures?

First things first: We do not claim that the age-related sensitivity to the agents of deaths in old trees or time-worn monkeys is to make a sacrifice to help the next generation of living trees or monkeys. Evolutionary biologists, if you say that, will accuse you of flailing in a quicksand of fundamental error. The deaths of elderly trees or monkeys at the end of their natural lives are not to "make room," as one might think from a cursory look at the pattern of species-specific senescence and life span. Senescence is instead generally seen as a by-product of life by the scientists who call themselves biogerontologists. But as we will see, senescence is still amazingly "tuned" to life, in particular, the "style" of life and the "style" of reproduction.

In the science of aging, the concept that senescence is an unfortunate by-product of some overwhelming benefits of some other characteristic that evolution has selected for is known as the "disposal soma" theory. *Soma* here refers simply to the body, not to the mythical drug of the ancient Aryans or in Aldous Huxley's novel *Brave New World*. The theory of the disposal soma revolves around the passive sacrifice of the body—its disposal—in the pursuit of successful reproduction. Reproduction is what is being selected for!

According to the disposal soma theory, after reproduction there is less and less selection pressure to keep the organism healthy. Aging is a by-product of this lack of selection pressure. Deleterious genes whose effects come later in life, after reproduction, can weaken the body, and yet the genes remain because there is no selection pressure to remove them. Also, genes that specifically aid efforts in reproduction, but also create havoc later in life, may persist through the generations. The best place to examine this evolutionary logic is in cases of sudden senescence, the extreme manifestation of the disposal soma theory.

To be sure, there is a spectrum in the varieties of senescence. The pattern of degradation varies from a gradual type of senescence to a more extreme senescence that is predictable in detail. And many viewers of the spectrum of senescence or nature lovers who seek out amazing dramas of evolution have focused on a particular example of awe-inspiring sudden demise. It involves a big, charismatic fish: the Pacific salmon.

All seven species of Pacific salmon exhibit sudden senescence. From eggs in an upland stream they hatch as fingerlings and spend time migrating downstream while feeding. They reach the ocean and live, typically, for two or three years out in the wild, open sea, growing large. When reproductively mature, they migrate back to the rivers and streams of their births to spawn. They journey upstream, swimming like Olympians in a grueling, grinding haul up rocky rapids, in a mass migration. Some turn color from brown to red. The male grows huge jaws for fighting with other males over the privilege of mating. Once upstream, the female deposits eggs in a streambed hollow she has formed with her tail or flippers, while the male fertilizes the eggs with sperm squirted on top of the eggs. After the mating period, females and males both die.

Measurements of the migrating, spawning salmon have shown elevated blood levels of a steroid and adrenal hormone called hydrocortisone. These levels damage the fish tissues in ways that resemble Cushing's disease in humans, an overactive adrenal syndrome. As a result, the salmon's liver, spleen, and kidneys degenerate. The upstream swimmers also stop eating, which wastes their bodies, and they undergo the general chemical stress of shifting from marine salt water, where they had lived for several years, to the freshwater stream for spawning. Finally, the stream migration itself is a physiological ordeal. The fish suffer abrasions when banging against rocks in fast-moving water. They take violent leaps to gain purchase toward the upland waters. Elevated steroids are apparently necessary biochemical helpers to propel the fish in their heroic feats. Thus a physiological behavior critical for species survival may also serve as an executioner. Life, in short, is traded for sex.

After those years of feeding and surviving in the ocean to build a large powerful fish, why haven't salmon evolved strong enough to endure the stresses of migration upstream to spawn and then to survive another journey to the ocean and back for future seasons of reproduction? The answer is not known. But biologists would generally agree that in the Pacific salmon we have an example of the body being disposed of in the further-ance of mating.

The idea is this: If an evolved physical or biological charac-teristic of an organism helps it reproduce and spread its genes to the next generation, and if that same characteristic inadver-tently slides the organism downhill toward death, the charac-teristic can establish itself in the population if the total increase in offspring now outweighs any increase in later offspring that would have occurred in the absence of that characteristic. This evolutionary trade-off is ultimately a quantitative game: What the Pacific salmon gained by evolving a particular way of reproducing in a single mating season must have outweighed an alternative lifestyle in which several years of mating yielded fewer baby salmon for their overall total output of offspring.

The disposal soma theory boils down to a question: Which provides the most spreading of genes—maintenance of the body for future reproduction or disposal of the body for a superior rate of reproduction now? Sudden, dramatic senes-cence is one pattern that answers this question.

Perhaps the most famously ephemeral insects are the mayflies. Adults of the more than two hundred genera of order Ephemeroptera emerge from their final underwater nymph stages into delicately winged forms ready for sex but not for eating. The mouths of mayflies are degenerate. It is as if evolution had designed them for doom.

Death in these dramatic cases appears to be programmed. Creatures who can't eat? Creatures who will die soon after mating? At an earlier mayfly stage of life, the scary-looking nymphs crawl around underwater as voracious predators with claw-like jaws and perfectly viable digestive systems. But as adults, their genes, which control their metamorphoses from the final nymph stages into the flight-ready adults, seem to have neglected to provide them with mouths. To add insult to injury, the adult abdomen could not digest any food that came its way and is usually inflated with air. Obviously the concerns in the evolution of the mayfly adult are flight and sex, not survival. These adults might only live minutes, or hours. If we feel at all sorry for their short adult phases, we should take note of their full aquatic, earlier lives, with often ten or more nymph stages, typically spread over several years.

Nature was not altogether cruel in giving birth to adult mayflies with dysfunctional guts and no mouths. Nature may have saved necessary resources for better wings, or more sperm and eggs, or more gyrating nuptial flights. We don't have the exact answers. The experiments have not been done to enable us to know precisely what body part was sacrificed so some other body part could be built that rendered better overall success in reproduction. But the logic has been laid out in evolutionary reality. Trade-offs have been made.

We have here a similar pattern of sudden senescence in Pacific salmon and in the mayflies. This is a case of convergent evolution—evolution in which several groups of animals independently discovered through the tinkering of evolutionary design that all-out sex that even entails death can be worth it in offspring produced. Many insects in fact share this pattern. As a variant of the theme, sometimes among insect species it is just the males that die after sex. What about other groups of species? No birds have yet been found that have catastrophic senescence following sex (probably because in birds both parents are necessary to rear the young). Intriguingly, in mammals, the pattern, though rare, is not totally absent.

In Australia live ten species of tiny marsupial mice in which quick death can be the wage of sex. Males are the ones who die more quickly, probably from stress. In lab colonies, males just under a year old undertake repeated sprees of copulation that can each last up to twelve hours before they collapse. Their deaths resemble that of the salmon. In the month before mating, the tiny marsupials' adrenals increase in size and the steroids in their blood rise fivefold. Other hormones elevate as well. If the males are captured before mating, and prevented from sexually dissipating themselves, they typically will live another three years, a huge increase in life span. Females survive for an additional year after their first litter, but the odds of their having a second are very low.

These marsupial mice, the Pacific salmon, and the mayflies all give a new meaning to the phrase *dying for sex*. I'm glad it's them and not me. For humans, evolution obviously and thankfully resulted in a different tuning of the life span. Our musical chords linger relatively long. But these cases of sudden, extreme senescence yield important lessons about the relationships among death, reproduction, and life, and they help clarify the essential evolutionary logic that will apply to the many guises of more gradual senescence.

9

Tuning Longevity

Is there a natural limit of human life? During the Italian Renaissance, Michelangelo lived to age 89. So did Epicurus almost two millennia earlier in ancient Greece. Though such enviously long lives were rare centuries ago, they show us something important: The longest-lived individuals of any historical era reached roughly the same age.

It is not the case that the most elderly Renaissance artists only reached 40 or that history produced a pack of wise ancient Greek philosophers who all died around 200. We must dismiss as hearsay biblical reports of amazing life spans, while the gods of mythology, of course, have always been postulated to possess great longevity. But real humans? Medical advances are pushing more of us into those once-rare stratospheres of longevity. Yet we can fairly well tell how old an elderly person is by looks. There is clearly a gene-based pattern, despite variation, to the average arc of human aging.

Except for those unfortunate few who bear certain rare diseases of the human maturation process, humans do not suffer extreme, crashing senescence like the mayflies or salmon. To what extent does some overall evolutionary logic apply across the spectrum of life spans of plant and animal kingdoms? (I leave out the fungi here, for simplicity.) We share a pattern with bristlecone pine, robin, beaver, chestnut oak—senescence is gradual. For these organisms, death does not follow rapidly on the heels of first sex.

Mammals, of course, need to raise their young after the

births. And mammals usually live to produce many rounds of future litters. Birds do, too. The same goes for many reptiles, amphibians, and fish. A lot of effort went into building their big bodies that started with tiny, fertilized single cells. Clearly it was worthwhile, in the evolutionary sense of reproductive success, to keep these large bodies around for a while. So what rules or patterns govern the various life spans?

We live long, but not as long as many trees, and not even as long as the oldest quahog clams, edible ocean bivalve mollusks that have been known to reach the admirable age of 220 years. A species of tortoise has a recorded life span of 150 years. Incredibly, these creatures not only beat us in years, they apparently do not appreciably senesce. The leading biogerontologist Caleb "Tuck" Finch has termed this phenomenon "negligible senescence." Other candidates for negligible senescence include several species of deep-sea fish. Individuals of the Northwest Pacific's rockfish and the Southern Hemisphere's orange roughy, for example, have been estimated to be 150 years old.

The point is not that creatures with negligible senescence live so much longer than humans—they might or might not. The issue is that in these species the individuals who are longest-lived (so far as we know) do not degrade from internal causes that would make their chances of death increase as they age. Some species with negligible senescence maintain high reproductive output despite their increasing years. In fact, among lobsters, another candidate species with negligible senescence, egg laying can become more copious with age.

Birds are turning out to be flashy mentors to the biogerontologists who patiently seek answers to deep questions about the spectrum of longevity in animals. Here's a perplexing finding that violates the general rule that large creatures tend to live longer than small creatures: A typical mouse of twenty grams lives about three years, while a canary of the same weight lives for twenty years, almost seven times as long. Limited data from wild birds show they live almost twice as long as same-weight mammals in captivity, while captive birds live about three times as long as captive mammals of the same weight. Some of the numbers are extraordinary. Scarlet macaws, for example, have been known to live more than ninety years, which is about four times the life span of average, similar-size birds and twelve times the mammal average at the same weight.

The explanation for avian longevity comes from what is known as the evolutionary theory of life span. Biogerontologist Steven Austad and his colleague Donna Holmes use the phrase *Fly now, die later* to describe it. The motto applies not just to birds but to bats, too. On average across a range of body weights, bats live about three times longer than other mammals of comparable weight. Austad and another colleague, Kathleen Fischer, hypothesize that the aerial abilities of birds and bats make them much less vulnerable to predators than are ground mammals. Austad and Fischer further reason that any mammal that can sail between trees should be better than ground dwellers at avoiding predation. They surveyed data for gliding species of mammals: three squirrels, five marsupials, and one flying "lemur." Taken together, these species have life spans that average 1.7 times the mammalian average for their weights. In another study, all marsupial mammals were lumped into two groups: tree-dwelling or ground-dwelling (species using both habitats were ignored). For comparable weights, the average life spans of arboreal species beat terrestrial ones by nearly 60 percent, Austad and Fischer found.

The core concept in the evolutionary theory of life spans is that creatures that are less vulnerable to predators are more likely to have evolved a healthy dose of maintenance and repair abilities inside their bodies. This is a central tenet of the overall evolutionary logic that relates life span to lifestyle: The intrinsic capacity of an animal's bodily metabolism to produce longevity is evolutionarily tuned to the odds that the animal will or will not be able to live long, on average, based on the relative kinds of advantages or disadvantages that its lifestyle confers on its survival.

Indeed, a number of lines of evidence suggest that the internal metabolic system of birds might be extraordinarily capable of self-repair. For example, experiments were conducted on liver cells of mice and three species of birds: parakeets, starlings, and canaries. All cells were subjected to a variety of stresses: 95 percent oxygen, which causes free radicals to form in the cells; hydrogen peroxide, a toxin produced by metabolism; paraquat, a chemical toxin; and gamma radiation, which causes genetic mutations as well as cell damage. These stresses cause cell populations from any animal to die off, but the cells from birds live much longer than those from the mice.

Already, scientists can begin to answer the question "Why do birds live so much longer than other animals of similar weight and size?" with some affirmative statements about enzyme repair deep inside their cells, though many details still need to be filled in. To understand the life span of birds more fully, we must let our attention sweep in scale from a bird's cells out to the whole bird; indeed, even to its relationship with its environment, its lifestyle.

In broad generality, birds live longer than mammals of the same weight because their airborne and arboreal lifestyles have made it economically feasible for them to develop the cellular mechanisms that enable them to live longer. The ultimate answer to their special longevity cannot be found inside their bodies, though; it is found in the fact that they are predation escape artists. Their relationship to sky and tree, where they wend their artistry, provides the causal impetus for the evolutionary honing inside their bodies of certain molecular dynamics, including their genetics. Thus, in the evolutionary theory of aging, the micro-level is correlated with, and in some sense is controlled by, the macro-level of what an organism's life habits are, how it avoids death in relationship to the environment and to other organisms.

The same reasoning applies to some of the deep-sea fish, and animals that have protective shells, such as the negligibly senescent quahog clam, tortoise, and lobster. Of course, many questions remain, because organisms have such complex relationships with their ecosystems of food and the dangers they face. But we can see some general trends, and recognize the truth that life span is no haphazard by-product of circumstance. Life span is an integral part of an animal's total relationship to its environment in evolutionary time, which measures trade-offs between reproductive success and metabolic maintenance.

What about humans? For one, researchers are studying an array of long-lived creatures with hopes of developing drugs that can transfer some of their metabolic tricks to us. Eventually, as I along with many other scientists believe, we will be able to start modifying our life spans not just by conquering diseases but also by manipulating systems such as that of insulin, which is proving to be a key to aging. But until those presumably happy days come to pass, how now should we regard the current natural human life span? To understand the central issues here, it's less fruitful to go into the details of, say, why Mr. So-and-So died at a ripe old age than it is to investigate why people grow old in the first place and what is the average human rate of aging.

Although humans are not one of the special creatures with negligible senescence, we should not have any species envy when it comes to longevity. We are definitely one of the exceptional cases. Compared with what would be predicted for mammals of our weight, humans live about four times longer. Some species of African antelopes, for example, are about our weight, and as average representatives for that weight class, live only about twenty years, max.

Could our longevity have something to do with the evolutionary process that turned other great apes into humans over millions of years? Thinking along lines of the advantages wings offer bats and birds, or shells offer clams, we should search for some special characteristic of humans that could have given our ancestors elevated skills in surviving the demolition derbies of predation and natural hazards, thus allowing the evolution of a long intrinsic life span to take advantage of that high survival rate. Was that characteristic our brains, our smarts? Or perhaps our ability to stand upright and run long distances during hunts (before the first big expansion of the brain in evolution)? Factor in our free hands that can hold stone tools to efficiently butcher game and even dig tubers. What happened over several million years that might have led to longer and longer life spans? And compared with which species? We should look at not only the mammalian average for our weight class but also the primate average.

As in groups such as birds or mammals, the subgroup of primates, considered as a set of data, shows a rough correlation between body weight and maximum life span. Those primates that weigh about a pound can live about as long as ten years. Tipping the scales at around ten pounds provides a life span potential of about twenty years. There is quite a bit of scatter, to be sure, but the correlation is good enough that a trend line—through the various data points on a graph of maximum life versus body weight—would predict that primates who weigh on average between one hundred and two hundred pounds should have a life span of about forty years. Some of our relatives weigh about that much. The chimpanzee weighs somewhat fewer than a hundred pounds, and the gorilla somewhat more than two hundred pounds. On the primate trend line of life span versus weight, both apes, in fact, fall very nearly on that forty-year prediction, with the lighter-weight chimp living a bit longer than the heavyweight gorilla.

So what about humans? In weight, on the average, we tote in between the chimp and the gorilla. Thus our predicted life span, if estimated from the average primate data line, should be in the neighborhood of forty years. But in fact our natural life span is more than double that.

(Note that this is true worldwide, if we consider the oldest within all cultures, including as well Renaissance Italy or ancient Greece. The concept here is maximum average natural life span, not actual average life span, because the latter factors in all kinds of special circumstances, such as childbirth complications, diseases, wars, and other external causes of death, as well as availability of quality medical care, all of which differ across cultures, both current and past.)

Among primates as a group, some of the boost in their average life spans per weight, compared with the mammal average, probably comes from their arboreal habits. "Climb now, die later," we might say. In trees, primates are out of the way of hyenas, wild dogs, and lions. Primates in trees can jump between limbs in ways that leopards, for example, cannot. Although primates can be eaten by other primates, such as monkeys that are killed and consumed by chimps, still, arboreal primates have been relatively free of predators. The average arboreal marsupial mammal, for example, lives nearly 60 percent longer than those marsupials stuck on the dangerous ground.

The rest of the boost in the primate life span likely comes from primate smarts. Graphing data for primates in terms of life span as a function of brain weight in fact gives a tighter correlation than it does as a function of body weight. The capacity for evasive behavior in creatures with large brains provides one route to escape from predation. Large brains also facilitate formation of complex social structures, with communication signals useful as warnings. If brain weight is taken to be the crucial factor, humans register as the top primate. And they possess the greatest longevity. Could those facts be closely related? I think so.

During human evolution we used tools far beyond the capacity of other primates. We had protective social structures. We probably used intelligence to review events by means of stories told by body gestures and then language, which aided learning and recall about danger, and about opportunities for the future. No one can yet definitely parse the details of this whole story. But it is likely that the two major pulses in brain size during human evolution over the past two million years were accompanied (or closely followed by) an increase in life span.

Our life spans derive from our place in nature, honed and changed through evolutionary time. For this we could be grateful to the smarts of our wily ancestors, because their activities brought about lifestyle conditions that facilitated the evolution of intrinsic better biochemistry for naturally longer lives. We live via the cumulative success of all those hominids, stretching back in deep time. In contemplating human biological death we might feel fortunate for our natural life span, and for all those who, over millions of years prior, made us what we are today.

The Awareness of Mortality

10

Death's Cultural Blooming

In *Grave Matters,* Nigel Barley has surveyed the astonishing (to me) customs of death around the world. He reports on his experience at the Indonesian island of Sulawesi with a Torajan granny who had been "sleeping" for three years in a home. The custom of wrapping the dead in "vast amounts of absorbent cloth to soak up the juices of putrefaction," while resources and money are being collected for the next stage of the funeral, sometimes leads to an extra piece of furniture for a while. Barley was requested to greet the granny in what "looked like a bundle of old clothes," which was being used as a shelf for cassette tapes.

Why are there customs surrounding death? They are clearly not direct products of biological evolution. We are genetically a single species with thousands of different customs. We invent practices that imbue death with meaning well beyond the fact of biological death. The rituals organize the survivors to gather and then to mourn, worship, recollect, sing, and act out solemn performances. The rituals are cultural inventions, products of a cultural evolution layered "on top" of biological fact.

As Barley's experience with the deceased granny shows, the corpse must be dealt with. Nonhuman animals, too, may mark the passing of one of their group. In one account of an adult female chimpanzee killed by a leopard, a dozen other chimps from her community gathered around her body. The powerful alpha male seemed to act as guard on the body for five hours, and allowed only the dead female's younger brother to groom the corpse and tug on her lifeless hand.

It is tempting to read into such observations the origins of human funerals and memorial services. When paleontologists use fossils and genetics to trace our lineage back in time, they find that around 6 or 7 million years ago it diverged from the lineage that gave rise to modern chimpanzees. If chimps reflect any ancient behaviors of the ancestor we shared with them, and there are reasons to suspect they do, the shared ancestor might have experienced emotional devastation at the death of a group member. But there is one big advantage the chimps have over us in dealing with the body. For a roaming band of chimps, the voracious biogeochemical recycling in the jungle—from carnivores to microbes—will quickly swallow up the dead.

At some point in human evolution we made a transition into a sedentary lifestyle, at least for portions of the seasonal cycle of the year. Caves were occupied by our ancestors, even before the Upper Paleolithic, which started roughly 40,000 years ago. Eventually these ancestors also made dwellings on the open plains. For example, excavations in Russia's late Upper Paleolithic have revealed domes made of mammoth bones. Neanderthals, too, who were not our direct ancestors but advanced hominids that shared descent with us from our ape and early genus *Homo* lineages, inhabited caves in Europe and in the Middle East.

A sedentary lifestyle, even when temporary, presented a particular challenge to a big-brained hominid confronted with death. We can imagine any number of situations. Lions fatally attack one hunter, but his companions manage to save the mangled body from the intended meal and drag it back to camp. Other deaths occur right in camp: A woman dies in childbirth, a baby born deformed lives for a while then dies, an injured forager makes it back to camp but then dies.

It would not be the brightest idea to leave dead human meat

in camp as an invitation to other carnivores. One option is to move the corpses well away from camp. These circumstances, born from necessities for survival—to avoid carnivores and limit exposure to possible diseases in the deceased—also provided the impetus for the invention of funeral rituals, or so one can easily imagine. Perhaps as early as a hundred thousand years ago, we might "see" groups walking in formation together as a few "pallbearers" bore the body a safe distance away. Perhaps sounds of lament, early music, rang out.

Skeptics to imputing ritual behavior from an archaeological record that only preserves inert objects suggest that the earliest actual burials in the occupied caves themselves might have been initiated primarily as predator lockouts. But Barbara King, anthropologist and theorist of the evolution of religion, points out that one extremely ancient Neanderthal burial involved much more than interring a body. At Regourdou cave in southern France a mind-numbing 65,000 years ago, a Neanderthal adult was buried with two leg bones and a split foreleg from a bear or bears, then covered with a slab of rock and a mixture of cobbles and an ash-like substance, all apparently then topped off and marked by an elk's antler. Other evidence of Neanderthal-gnawed bear bones in a purposively constructed nearby cache indicates a communal meal, possibly as part of the "ceremony."

King shows, starting with primate evidence, that there will be deep emotional connections among the individuals in such groups. As studies with the chimpanzees confirm, individuals know one another and even know about relationships among the members. According to King's thesis, we are witnessing in the earliest evidence of death rituals, such as Regourdou cave, a profoundly developed level of "belongingness," which represents the social evolution of the group as an entity bound by an intense degree of need, respect, even worship among its members.

Today, when someone departs this earthly coil, family and friends who have sometimes not seen one another for years come together. Often they are psychologically wounded. Suddenly the social network that constituted the deceased's life congeals into a poignant focus. But the deceased is missing. The effects of death are wounds to that social body, tears in its fabric.

For the ongoing integrity of the living body, wounds must be healed, tears repaired. In response to biological cuts and blows, the immune system kicks in. Special cells rush to the site of the rift and start the repair, inducing neighboring cells to grow rapidly and fill in. It's amazing to watch from a distance, as you undoubtedly have done with a cut on your skin. How do the cells know what to do? How do people know what to do when faced with a death in the family? Special "cells" are sent in—the police, coroners, lawyers, priests, rabbis, Buddhist monks, or shamans. Cultural evolution has produced, through many thousands of years of honing, ways of helping the social cells in need of healing. The wounded also rely for solace on their own feelings and instincts. Touches of the personal may be combined with well-honed wisdom. At my friend John's memorial, for example, his dulcimer was displayed front and center, and on it hung a small plaque that read, DANCE AS IF NO ONE WERE WATCHING, SING AS IF NO ONE WERE LISTENING, AND LIVE EACH DAY AS IF IT WERE YOUR LAST.

It seems reasonable to consider ceremonies surrounding death as the earliest of all rituals in our species' history. Compare death to the other universal moments of transition that have also been ritualized, such as birth, puberty, and marriage. Many babies in human prehistory would not have lived long, so there is no good reason to think that birth rituals preceded death rituals in cultural evolution. Marriage could have been ancient, but because it is more abstract than death, marriage as a ritual probably came to the human scene later. Puberty rites were not immediately necessary: Biological puberty happened anyway.

The archaeological record contains evidence of a number of burials, both of Neanderthal and of our own species. One remarkable early ceremony of *Homo sapiens* took place in what is today's Russia. At a site called Sungir, archaeologists have uncovered evidence of extraordinary rituals 28,000 years ago.

In the Sungir burials, grave goods were interred in an array with the bodies of an elderly man and an adolescent boy and younger girl. Across their torsos and heads were strands of more than ten thousand beads of mammoth ivory, apparently once attached to clothing and caps, now long gone. Several hundred teeth of polar fox remained in the region of what had been a decorated belt for the lad. The burials also held pins of ivory, ivory carvings that included that of a large mammoth, ivory lances and disks with holes, a schist pendant painted red, antler batons, and a polished human femur bone.

Twenty-eight thousand years is a long time ago, indeed, thousands of years before even the coldest time of the last ice age. Yet it is not so difficult to imagine us there at the burial, sympathizing with those relatives of the Upper Paleolithic. Given such care in outfitting the dead with beads and other valuable objects, living memories of the deceased would have been consolidated and amplified. Such ritualized handling of the dead and visual splendor in the burial art could have induced the living to contemplate death, thus providing the survivors with imagery about their own futures and about the mysterious flow of life. By this era, with special burials with such elaborate burial goods, death rituals might even have fulfilled the function of seeming to help the dead. In the words of Steven Mithen, respected British archaeologist and author of *The Singing Neanderthals* and *The Prehistory of the Mind,* "It is difficult to believe that such investment would have been made in burial ritual, as at Sungir, had there been no concept of death as a transition to a non-physical form." Barbara King suggests a similar interpretation is at least possible in the even earlier Neanderthal burial at Regourdou cave.

Now fast-forward many thousands of years. Around ten millennia ago, early agriculturalists in the eastern Mediterranean region practiced mortuary cults in which the skulls of dead ancestors were fleshed out with clay. Closely following that, in the village of Çatalhöyük in today's central Turkey, bodies were buried in the clay floors of mud and plaster dwellings. The narrative focus of the world's earliest surviving recorded epic tells of how King Gilgamesh is shocked into an awareness of his own mortality by the death of his friend Enkido. Roughly at the same time, four to five millennia ago, giant pyramid-tombs, the largest surviving structures from the ancient world, were built by the Old Kingdom of Egypt. During the third century BCE, the "First Emperor" of China not only vastly extended the Great Wall but spread an expanding wall of paranoia about his eventual death, sponsoring expeditions to seek for him an elixir of immortality. His bodily burial included "armies" of buried life-size clay figures. We still do not know if deep inside the unexcavated giant earth mound that contains his tomb is the mythically reported lake of mercury, which was one suspected element in his elixir for immortality.

Fashions in rituals of disposing of and saying good-bye to the dead have come and gone. With the advent of Christianity and other cultural shifts, the ancient customs of burying grave goods with the deceased, except for clothing, have waned in substantial portions of the world. Cremation has often oscillated with burial as the favored death ritual. Today new practices are being designed and implemented. The objective of "green" burials is to allow the body to quickly return to soil via worms and microbes but under special trees.

We might look at the effect of death on the cultural aspects of life almost like that of a small particle of rock being dropped into a supersaturated solution of some mineral ion. The particle acts as a seed to start a crystal rapidly growing. In a society, consider the people out and about doing their business like ions of dissolved minerals in solution. They are supersaturated: All exist in a state of potential upset, as news could reach them about the death of a friend or relative, or they themselves could die at any moment. Suddenly an actual death plops into reality. And immediately the ions crystallize around the dead person in organized rituals. Each ritual is true to its culture, as crystals of calcium or silicon will be true to their form, given their own specific chemical characteristics. But as a universal pattern, social life thereby forms around individual death.

The manner in which such memories are institutionalized affects the living in more ways than just specific recollections of the dead. The living witness not only what is remembered, but also the tone of the funeral. Worldwide, the most universal emotional tones are grief and respect. When the living see that others lament the dead, they are consoled about their own future deaths. Why should anyone, at least those who do not believe in an afterlife, care how they are remembered? The concern is widespread and seems almost instinctual. Indeed, the prospect of one's own death that comes up in daily life via ordinary human awareness is fearful enough without seeing someone dead. But when such trauma occurs, the fear is elevated and then assuaged by respectful burial or memorial practices. We learn gratitude for our relationships with others from such ceremonies and anticipate that others will similarly feel gratitude toward us when we die.

I am reminded of a joke. A comedian first bemuses about the fact that we are taught only to speak good of the dead. "What's that you say? *He's* dead? Well, good!"

What makes this funny is that we probably want to say it more often than we dare (fill in your least favorite politician here). Very improper, to say the least. It is also funny because it violates basic instincts and puts us in touch with our fears. It would be fearful to live in a society with such attitudes about the dead as the norm.

That the dead cannot be treated with the disdain in our imaginary experiment is quite obvious within a modern, stable society, you might say. True. But only because such treatment is abhorrent and unimaginable (war is another matter). Such a dug-in feeling reveals something profound. It shows how connected we are, how dependent we are on existence in the thoughts of others. Death can explosively bring to the mind's surface such dependency. Somewhere in the human mind, not as cool calculation but as boiling emotion, is a golden rule of death: Do unto dead others as you would like to have done unto dead you.

This is rich material, with ties to the variety of cultural beliefs about life and death. But what most impresses me is the overall constancy of it all. In the time line from the very ancient ice-age cave burials to the present we find much that is consistent. Death, though it might be nothing to the deceased, meant a lot to the living. Men and women of power, fortune, or fame tended then, just as now, to have bigger wakes made in the wake of their deaths. Evolutionary psychologists claim that down deep in the cognitive processes of our brains and in the basic dynamics of what culture has provided for our psychological needs since ancient times, we have not changed all that much. With the advent of cultural evolution, the tie between death and life took a new twist, with the death of others ever since affecting the behaviors and psychologies of the living.

Death-Denying Defenses

What happens when people are reminded that they eventually are going to die, that they are mortal? Epicurus and Buddha encouraged their followers to remind themselves, as constant meditation. Thinking about mortality, they and many others have claimed, is a path to deep contentment, even happiness.

But do we need to remind ourselves? The world we live in seems to do a great job of that already. We are exposed to the fact of mortality all the time—movies (what's a movie without death or near-death?), news ("if it bleeds, it leads"), the health problems of family and friends. Sometimes living within a death-soaked world can be disturbing. But after all, it is the state of nature with food webs and biogeochemical cycles. Almost inevitably, this state has been stepped up a notch by the ability of humans to symbolize, fantasize, and hypothesize, not to mention our recognition of real murder and genocide. What more can be known? In recent years some experiments in the field of social psychology called "terror management theory" have been yielding some fascinating insights on how the basic knowledge of our mortality affects us.

If you happen to be a student in an introductory psychology course at a college or university, you are usually required to participate in one or more experiments. Undergraduate psychology students are needed to form a large enough pool of "lab rats" to provide statistically relevant results. If your school was one that had experiments in terror management theory (among a large number of other types of experiments), you might find yourself in a room at a desk filling out a "personality questionnaire." At least, that is what you are told.

You are told nothing of the true purpose of the overall experiment. Instead, you might be informed that the questionnaire is about personality traits and interpersonal judgments. Scattered among a range of questions to answer are the following: "Please briefly describe the emotions that the thought of your own death arouses in you," and "Jot down, as specifically as you can, what you think will happen to you when you die and once you are physically dead."

As you write short answers to these questions, you don't know that only half of the subjects are filling out a questionnaire structured exactly like yours. For the other half of the group, the "death" questions have been replaced by others. The real goal of the questionnaire, which you randomly received, is to remind you, with a slight prick to consciousness, of the fact of your mortality. For you, death has been made briefly tangible. You write and pass on to other questions, and then to the next segment of the experiment designed to answer the research question, Will that death reminder affect how you respond to threats to your cultural worldviews?

The original three leaders of terror management theory were Jeff Greenberg of the University of Arizona in Tucson, Sheldon Solomon at Skidmore College in New York State, and Tom Pyszczynski of the University of Colorado in Colorado Springs, all professors of social psychology. Many others have since joined in what has become an international endeavor. Before describing the results of this research, it is helpful to know something of the impetus for original researchers' thinking.

Terror management theory starts with the straightforward observation that humans have an urge to live. There is a joke about a prisoner on death row. As he is about to be executed, he is asked if he has any final requests. He begs permission to sing his favorite song, all the way through. Go right ahead, was the reply. The prisoner begins: "A million bottles of beer on the wall . . ."

As we have seen, evolution has already tuned us for relatively long lives, compared with our great ape relatives of about the same body mass. But we want more. We are survival-seeking machines who now read articles and talk to others for clues about living longer and healthier lives. Except for the unfortunate cases of depressive suicide, the search for longevity is natural and instinctive. A recent extreme extension of this drive for longevity is the idea that we will be able to overcome normal biological senescence and death by becoming machine–human cyborgs or by downloading consciousness into advanced computer software. Others of us might manifest the drive simply by seeking to "go organic," to exercise regularly, to make love more often. To me, it all seems part of our deep endowment to want to live.

Yet in the words of the terror management researchers, there is also a "worm at the core." Perhaps this psychic worm was already present at the time of the Neanderthal and *Homo sapiens* burials during the last ice age, complete with grave goods. Perhaps an ancestor of yours, a participant who took part in the hypothetical funeral procession to move a corpse away from camp a hundred or more millennia ago, looked into a pool of water and felt the wriggling presence of the worm. We will never know the origins of the moment. But certainly, somewhere in human evolution, a simple, symbolic, thinking process of our brains, which was so helpful for survival and based to a large measure on the ability to make projections into the future, started to realize the horrible truth—not only do other humans die, but "I," being human, will also die.

The realization of one's inevitable death, born from a magnificent brain, conflicts with the functional purpose that brain has to survive. Today we confront this fact even more poignantly than ever. We witness medical advances that often extend life immensely—but not indefinitely. There exists within us a deep conflict between our will to survive and our knowledge that the thrust to survive will ultimately be thwarted. Call this collision the "primal dilemma."

The researchers of terror management were originally inspired by the work of psychologist Ernst Becker, who won a Pulitzer Prize in 1974 for his book *The Denial of Death*. According to terror management theory and Becker, the simplest ways that our fear is controlled and overridden are through employment of what are called "immortality projects."

Some immortality projects are literal. In them actual immortality is promised, and thereby a premise in the primal dilemma is negated. In this category we find the many types of afterlife scenarios, projected by different cultures throughout history, some of which are still very active today. The promise of immortality occurs, for example, in the final line of a bedtime prayer I was taught in childhood:

> *Now I lay me down to sleep,*
> *I pray the Lord my soul to keep;*
> *If I should die before I wake,*
> *I pray the Lord my soul to take.*

Other immortality projects may be purely symbolic. These operate through identifying and seeking ongoing life in one's words, or the spread of one's personality and beliefs. Seeing your progeny carry on, or writing books that you hope will outlast you, for example, are forms of symbolic immortality projects. When I walked in the New Mexico evening deeply fearful about my health, I comforted myself by mulling over all my connections to others—an instance of experiencing an "extended self," an immortality project of a kind.

Beyond such relatively conscious immortality projects, the terror management researchers have been probing what might be called the immortality projects of our subconscious. They ask, What do we do about the primal dilemma on a day-to-day, a moment-to-moment basis? We might forget about it but the worm is still there, buried and unseen. And perhaps working away within us . . .

In the next phase of the experiment, whether or not you are one of the subjects who have been pricked with awareness of mortality by the questions "buried" in the questionnaire, you are presented with essays about America. These are purportedly authored by foreign students studying in the United States and "published" in a political science journal, but they were actually written by the psychologists and the journal is bogus. Subjects are asked to read these essays and then rate how much they like or dislike the authors. One essay is heavily pro–United States in bias, with the rah-rah message that the US is the greatest nation, the land of milk and honey, etcetera. The other essay is bitterly anti-US in orientation, claiming that the nation's ideals are no more than advertising facade while the reality is a nation in which the rich become ever wealthier and the poor more downtrodden.

Here are the key findings: In a typical experiment, subjects in both groups (those who were pricked with death awareness and those who were not), on average, judge the pro-US essayist as more likable than the anti-US writer. However, the subjects whose potential for inner terror had presumably been perturbed by questions about their own deaths ranked the pro-US essayist as exceedingly likable and the anti-US essayist as exceedingly disagreeable compared with the rankings given by the control subjects.

According to the theory, the subjects who are subtly reminded of their mortality will want to bolster their inner defenses against their inner terror, and they do so by defending their "cultural worldview." What are cultural worldviews? They are "humanly created and transmitted beliefs about the nature of reality shared by groups of individuals." They are the brick-work in our edifices of politics, religion, economics, law, sports, fashion, and so forth. Some examples of cultural worldviews: that Catholics have the correct take on the nature of God, that hair dyed green is the ultimate in cool, that the spread of capi-talism is the best way to upgrade living standards around the world, that the albatross is the most critical seabird to conserve right now.

There is often a fuzzy border between a fact and a cultural worldview. That DNA has four kinds of bases is obviously a fact and not a worldview. But DNA can become an object within a worldview when people believe, say, that DNA analysis is the royal road to a terrific human future. Note that cultural worldviews are, by definition, not strictly private. They are worldviews because they are shared by many people—but not everyone. This creates what sociologists call an in-group of those with the shared view, while others constitute the out-group.

Holding cultural worldviews seems like the ordinary bread of everyday, everyman-and-everywoman human existence. And so they are. But terror management theorists take this a step farther to ask, Why do we hold worldviews, and some-times so tenaciously, even in the face of contrary facts? The amazing advance that the terror management psychologists have made is to show us that this quintessentially human attri-bute at least partially, and perhaps largely, functions to help us live with the primal dilemma. The mechanism operates so smoothly as a system of psychological defense against the terror of death that we do not even suspect it functions as such.

Cultural worldviews ameliorate anxiety of vulnerability and death, according to these social psychologists, "by imbuing the universe with order and meaning, by providing standards of value that are derived from that meaningful conception of reality, and by promising protection and death transcendence to those who meet those standards of value." To the extent this is true, the rationale for the design of the experiments and the experimental results neatly follow. If the construction and maintenance of worldviews protects us from the disturbances of the primal dilemma, then bringing awareness of mortality, even briefly, into consciousness should cause us to intensify that construction and maintenance.

This is what happens in the experiments with the bogus essays. The death-aware subjects praised more vociferously than did the controls those who seemed to be similar to themselves. And they denigrated those who held different opinions than the mainstream values of their own worldview as US citizens. It is important to note that the subjects do not necessarily feel increased terror consciously, and then seek to relieve it by defending their worldview. The idea of terror management is more subtle. The subjects, like everyone, live immersed within the invisible, unconscious dynamics of daily life, which usually creates a mildly (at least) pleasant existence, continuously suppressing the disturbing idea of death. If death awareness is awakened just a bit by the death items in the questionnaire, you are not even aware of its awakening. Nor are you aware of how the worldview defense mechanism kicks in more forcefully than normal. This all happens in the unconscious.

How do we know it was actually the idea of death and not a general anxiety that caused the responsive defense of worldview? In other words, was the more extreme evaluation of the essayists by the test group of students due to their state of anxiety? Could any garden-variety anxiety produce the worldview defense?

To examine this issue, the researchers have administered an additional questionnaire designed to measure general anxiety, after the introductory personality questionnaire but before students read and evaluated the essays. Significantly, the subjects who were pricked by the knowledge of mortality did not exhibit elevated levels of anxiety. Furthermore, questions about final exams and other issues that were proven by the research to provoke anxiety failed to produce the worldview defense. This suggests that anxiety itself does not trigger the reaction of worldview defense. But being reminded of one's mortality does.

These essential findings have been replicated and solidified by countering other alternative explanations proffered over the years. In hundreds of experiments in both the lab and out in the field, and in countries such as Canada, Germany, and Israel, as well as in the United States, the theory has stood up, and has become further elaborated as findings from different experiments start to form the coherent whole of an overarching theory. For example, some of the field tests involved questioning pedestrians about a hot political issue—immigration laws, in one German experiment—which then showed that responders exhibited more worldview defense when the interview took place in front of a funeral parlor, rather than a distance away.

The psychologists also tested actual Tucson judges who volunteered for an experiment. The judges were given legal briefs of arrests for charges of prostitution, and were then asked what bail they would normally set in such cases. Half had been exposed to the death questions in the "personality" questionnaire and half were controls who had answered the same questionnaire except for the death questions. Bail set by the death-pricked judges was nine times higher. In yet other experiments, students who had undergone death awareness bestowed higher rewards for heroes. This showed that the defense mechanism was not necessarily tied to punishing those in one's out-group. It could operate as well by supporting those in one's in-group.

In some particularly fascinating work, the psychologists have demonstrated the importance of the unconscious in these dynamics. For example, if the test subjects are instructed to think very carefully about why they are passing judgment, say on the pro-US and anti-US essayists, the distorted ranking toward in-group and against out-group does not occur. Making the process conscious, this suggests, does not allow the unconscious to perform the repression of the specific disturbances by the seemingly irrational mechanism of worldview defense.

The power of the unconscious is further shown by the fact that the induction of death awareness itself does not have to be conscious, even for that brief period of time required to answer the two questions about death on the original questionnaire. In one set of trials, the psychologists flashed on a screen mortality-reminding words such as *death* or *dead* too quickly to be consciously registered, and sandwiched those between neutral words that were on the screen long enough to be readable. Despite their invisibility, worldview defense was still brought on by the subliminal mortality-reminding words. The words must have been registered somewhere in the brain by the unconscious, which subsequently cranked up its worldview defense mechanism. Ah, but what if another potentially disturbing word such as *pain* were to be flashed as the sandwiched trigger? This was tried, and the results were negative, as predicted by the theory. In the experiments, only death reminders seem to spark the worldview defense!

The experiments provide modern rationale to why way back in the time of Epicurus, more than two millennia ago, it was recognized that the path to psychological calm by accepting that "death is nothing" was arduous. Incorporating into our deepest being that ancient Greek logic—"while we exist, death is not present, and whenever death is present, we do not exist"—requires effort, because death is present, in a real sense, in our inescapable inner awareness of mortality. The discoveries by the social psychologists of the consequences of this awareness reveal how profoundly it influences and even shapes our being.

The implications for us: Worldview defense is a specific antidote against psychological disturbance engendered by awareness of one's mortality, and the defense is so well constructed that under normal circumstances the disturbance is controlled to a level that we do not even notice. Before existential terror or other conscious problems arise from any trigger of the awareness of mortality, worldview defense kicks into action. And the process appears to be unconscious.

12

Death and Self-Esteem

One way we attempt to cope with the thought of our own death, an impressive amount of research suggests, is by trying to enhance our self-esteem. Sigmund Freud captured a dimension of this impulse concisely when he remarked that "To the writer immortality evidently means being loved by any number of anonymous people."

Some of the experiments on how enhanced self-esteem can buffer death anxiety have used neat variations on the well-established experiments on worldview defense. The basic idea behind this research is that if either self-esteem enhancement or worldview defense can function to repress the disturbance that might arise from death awareness, then subjects who are given boosts to their self-esteem should have less need to defend their worldview. This is indeed what has been found.

For example, the self-esteem of subjects can be manipulated early in the experiment by having them undertake a word puzzle. No matter what their actual accomplishment, they can be told that they scored either in the upper echelon or at the bottom rung. Those that were told they were top dogs sally forth into the next phase of the experiment with enhanced self-esteem; the others have their tails hanging in shame. Now the experiment continues with the questionnaire that includes death-awareness topics and then, finally, for instance, with reading and ranking of the pro-US and anti-US essays. Only the subjects with lowered self-esteem gave extreme reactions to the essays; their knowledge of apparent low ranking caused them to become more defensive about their own worldview. On the other hand, subjects with artificially induced, high-flying self-esteem merely ranked the essays the same as a control group did.

Thus a high level of self-esteem by itself functions as an anxiety buffer with no apparent need to call on additional buffering using worldview defense. Self-esteem and cultural worldviews are clearly related. With cultural worldviews we see ourselves as part of a larger shared system of specific beliefs and values. With self-esteem we see ourselves as a valued member of the in-group with a specific worldview. The experiments with self-esteem can also be conducted with a focus on the human body. If there's one thing all of us know about death, it's that our bodies age and undergo eventual, terminal demise. Being reminded of one's body can be a source of lowered self-esteem, since awareness of our raw, animal presence is a message about mortality. As animals, we know ourselves as material entities that ultimately age and decay.

This particular twist in the theory was tested by social psychologist Jamie Arndt of the University of Missouri. All the test subjects wrote responses to the personality questionnaire with the standard two death questions. Thus everyone was primed with death awareness. Now, could awareness of their bodies add or subtract from this priming? The students wrote in small, private rooms. Half the students composed their answers at a desk that had a large mirror in front. These subjects could not avoid at least occasionally glimpsing themselves in the mirror. The other half wrote in rooms that were identical in all ways except without mirrors.

The theory of terror management suggests that because self-awareness is necessary for the idea that we exist, it is also necessary for the experience of the primal dilemma—the clash between the need for life and the knowledge that death will thwart this need—which requires the concept that someday we will not exist. Being faced (literally) with one's physicality is a worrisome threat to dignity. Therefore heightened self-awareness and feedback on one's physicality should increase the possibility of reflecting, even if unconsciously, on the inevitability of one's death and thus feeling the terror of the primal dilemma. What Arndt measured was something very simple: How long did the subjects take to fill out the questionnaires that included the two death-related questions? Did the mirror make any difference?

According to the theory, subjects in rooms with mirrors, in facing questions about death, would be less able to face themselves in the mirrors. So they would more quickly write and thus more quickly exit the room.

The specific results? Subjects who sat in the rooms with no mirrors took a little over six and a half minutes to complete the questionnaire. But subjects who sat in rooms with the mirror took nearly two minutes less. Similar results have been observed in other trials designed to test whether general anxiety alone (say, generated by questions about final exams or future careers, without mention of death) was the operative factor. Again, only the contemplation of death produced the difference in timing between the cases with or without mirrors.

What are the implications of all these remarkable findings of terror management theory? Death awareness doesn't only affect rituals that bind culture. Death awareness also influences—and thus helps to give life to—the daily, pervasive mental structures by which we are bound together with shared worldviews and methods of supporting one another's self-esteem. Complex aspects of our psyches, aspects that might be thought independent from the concept of death, are at least partly and perhaps substantially forged by the presence of the worm of knowledge about mortality. Our very psychological life and the way we share reality with others is partially sculpted as a response to death. As biological death is turned eventually into new life forms via carbon flows in the biosphere, so the cognitive aspects of human death are turned into social and psychological structures via interrelations between individuals and their cultures.

The theory of terror management may also provide us with new insights into the origin and maintenance of human aggression. An ultimate way to defend one's worldview is simply to eliminate the others, thereby proving that "my" worldview must be the correct one. If you can't persuade the out-group to join you, in other words, then annihilate it. In one variant of the terror management experiments, Sheldon Solomon used a killer hot sauce as "punishment." Subjects first filled out a standard battery of terror management questionnaires, and then they moved on to what they were told was a second complete experiment. There, they were to allot an amount of the hot sauce for a taste test to a participant from the "first" experiment with either similar or dissimilar political views. Those who had gone through the priming of death awareness "punished" with double the amount of hot sauce those with dissimilar views, compared with the amounts they allotted to those of similar political persuasion.

These results, and the theory behind them, indicate the possibility that the aggressiveness of wars and other conflicts (perhaps in day-to-day business as well) is in part a function of disturbance over mortality, which can come to life whenever differences in worldview become known. These differences threaten the respective worldviews, a process that thereby decreases the level of psychic protection they provide and allows the terrifying knowledge of mortality to rise up.

Because everyone holds a variety of cultural worldviews and is faced with the fact that many others may not hold those views (or may often support contrary views), we experience threats to our defense systems all the time. Thus we are tempted to denigrate. One aid for this situation has come from the experiments. Tolerance itself can be a worldview. When the psychologists took subjects with extremely liberal values through the battery of tests of terror management, the subjects did not need to denigrate their worldview rivals—conservatives, for example. In fact, after mortality salience, liberals liked conservatives more, presumably because they were in essence defending their own value of liberality by putting it into action in becoming even more tolerant.

Thus the theory of terror management gives educators and social scientists the means to study tolerance in a new way. We can do harm by our opinions that are not true judgments but prejudices, powered by the armor they provide against knowledge of our own death. If we can become aware of this process, death as a motivator in our unconscious could be put to use, theoretically, to create a more just and tolerant society. The energy in wood can be used to burn down a town or to power a steam engine and make electricity. In other words, we need not be helpless in the face of the psyche-sculpting power of death awareness, but rather can use the power of that knowledge to encourage socially desirable reactions in individuals.

To repeat: The tolerant can be more tolerant when faced with the awareness of death. The bigoted become more bigoted. It's a choice we need to think about. We should study how to develop self-esteem in ways that include everyone, and we might eventually be able to harness the power of death awareness to this regard. Indeed, we might have little choice but to use death awareness as an important component in our quest to enhance life and make a more desirable society.

An example from my own experience may give some sense of how these provocative questions can lead to fresh insights. For several days after an intense round of reading papers on terror management, I found myself reviewing the memory of an event from my childhood. I was about thirteen years old and at the time reading one of the volumes in the Time Life Science Library. I loved those books because they held fascinating text and sumptuous pictures. They were rich fertilizer for my brain. This remembered book was in fact about the brain. A section about memory dealt with intelligence and a metric of intelligence called the intelligence quotient, IQ.

The illustrations showed pictures of famous people, all now long dead. Along with their images, IQs were printed in bold. Their IQs had not been measured with an elaborate version of the twentieth century's Stanford-Binet test, which I had taken a couple of years earlier for possible entry into a college preparatory school. Rather, psychologists had estimated (obviously correctly, in the mind of a thirteen-year-old) the scores of these famous personages. Included were people whose lives were familiar to me, such as George Washington. Others I had never heard of. Already having learned the results of my IQ test, I was overjoyed to see that my score was higher than that of Washington himself. *Wow,* I thought. That meant I would probably go down in history too and someday have my name and picture in a book such as this!

True, my rank paled before that of someone named Voltaire. But who was he, anyway? And what about this guy Goethe, who also boasted a number well above mine? Well, whoever they were, my being superior in IQ to Washington was more than good enough for me.

What security I gained and what self-esteem I enhanced by seeing that my brain's capabilities were situated within the range of those on that fateful page. What immortality I then saw ahead for myself. All that was needed was for me to do something famous. But that was, as I felt then, basically in the bag, or in the brain.

It has become surprisingly clear to me that ever since I was a young teen I have been using this incident, and others like it, as part of my personal terror management. And, further, I now realize that one of my goals in life has been fame. It was not always conscious, but even unconscious it was a deeply seated motivation. I believe I was using it as a way of grasping at immortality in the face of obvious mortality. These high-IQ guys took hold of my mind as I gazed at their mugs in that book. Examination of death and the mechanisms by which we incorporate death awareness into our own lives might lead, as it did in my case, to surprising realizations far afield from the last sting we felt at the loss of an admired celebrity, friend, or relative.

Because many of our psychological structures that deal with the primal dilemma operate unconsciously, the life that death awareness helps to create might not automatically be the most desirable. But knowledge of how we attempt to ward off thoughts of death, and what a hold that avoidance can have on us, gives us more conscious choices as we watch ourselves search for self-esteem and embrace particular cultural world-views. It might be best, for example, not to cling to certain desires that are, in part, reactions to death awareness, desires such as that for, in my case, fame.

The findings from terror management theory can them-selves become a kind of culture we can learn to live within, a kind of meta-culture, or a worldview about worldviews. And such "going beyond" can lead to liberation through knowl-edge, knowledge that the awareness of death drives the society and its individuals in ways scarcely perceived. Why liberation? The path is yet to be determined, but we can be sure that with new knowledge of the shadow of death comes new potential for life.

13

Ripples upon Flowing Water

I vividly recall my first visit in London to Westminster Abbey. I read before entering that within I would see tombs of kings and queens and many other notables. I had gone to see the spacious architecture; as for those decorated tombs and monuments, well, I thought I could miss them. Yet within moments I forgot the building and became entranced by the iconic records of past people. I found myself wandering around in an almost drunken, mystical state in which the living and the dead were all woven together into one barely imaginable, integrated, and global, psychological system.

The experience was perhaps closest to what I had felt the first time I walked through Muir Woods, the magical valley just north of San Francisco. There I beheld the majestic, behemoth trees tightly integrated to form the so-called cathedral redwood forest. Just as Muir Woods is imbued with the sense of an extended organism, so the abbey holds the experience of the extended self—the integral larger psychological being of the social world. Perhaps the experience of interconnection on two scales can be an answer to the issues raised by the revelations of research on terror management.

In the abbey the web of history seems simply to seep from tombs of ancient kings and queens. You can sense the intrigue, perseverance, ambition, and pace of their full lives, even if you are not a historian. But inspiration also comes in the lesser-visited corners and tombs in walls. Words of wisdom, such as MANS LIFE IS MEASURED BY THE WORKE NOT DAYES, are so much weightier carved in stone than rolled out as mere ink on printed page. Along the north aisle of the nave, on a plaque near the body of a Mrs. Mary Beaufoy, you can read:

> READER, WHO ERE THOU ART, LET THE FRIGHT OF THIS
> TOMBE IMPRINT IN THY MIND, THAT YOUNG AND OLD
> (WITHOUT DISTINCTION) LEAVE THIS WORLD, AND
> THEREFORE FAIL NOT TO SECURE THE NEXT.

What a perfect example of terror management! First comes the terror, "the fright of this Tombe," then relief, "fail not to secure the next [World]." Such a gem from a past psyche, however, is not the main treat of the abbey.

The abbey's true gift is its three-dimensional immersion in the extended self. You can find this in any large, vibrant city, of course, but not with the inextricably integrated dead to boot. In the abbey the unavoidable dead reach right into the living. They are people who have set in motion the ripples of history, and thereby our lives today. Kings, queens, scientists, poets, thinkers, musicians. Some you probably know well, others not at all. So you may not be conscious of how they each have wrought their effects. That doesn't matter. They have affected somebody, and these "somebodies" were alive in the world and have affected others who have affected you and me. Sometimes the most profound effects originate with individuals you don't even know existed. Who influenced the course of the history of music? Or that of the written word? Many in the abbey and many, many more elsewhere all over the world. What about politics? Political structures ripple down to us from past leaders in such a manner that only the astute expert could trace how a twelfth-century decision evolved and thus lives on in the world today.

Now consider again the valley of giant redwoods. A single tree is easy to focus upon, perhaps to study its height or to gain a detailed look at its rust-colored bark. But what about Muir Woods itself? First of all, it must be experienced in time and space, with a hike. Then it can be conceptualized by memory of the hike. But if one tree is overwhelming, how much more so the forest! The situation is similar in Westminster Abbey. Stand before a monument to one poet, say Shakespeare, and a normal state of awe can be fairly well maintained. But take in the whole Poets' Corner, wherein lie the bodies unheralded by most today, and one begins to reel from the "forest" of talent and recognition of the outpouring of collective, literary humanity, a woods too vast to be encompassed by any one living individual.

The ecological metaphor seems to serve well for the abbey experience. The organism in nature is like the person in society. As one noses into some small alcove just to delight in what statue or plaque it may hold, exploring as one might explore nature on a sunny day too glorious to plan each step, one asks: *Who was this person? I don't even recognize the name, yet the former life is palpable.* Similarly, in nature, *Who is this insect hidden in the folds of the purple thistle flower? I don't know its name.* Freely investigating details, in either place, can produce a pleasant buzz in perceiving the sublime complexity of individuals in a larger whole.

Let's push the analogy farther. *Where in the ecosystem are the dead?* For that, look down to where the worms feed, where roots probe and microbes rule: the soil. Into that dark, moist realm fall the needles, the leaves, the feces, the corpses. Certainly this is the place in nature analogous to what the abbey represents. In the soil are products from creatures past. Litter in the top zone is just a year or two old. Beneath is former life now aged to decades-old detritus. Farther down the time scale, the age of particles found approaches centuries. In the abbey we sense the soil that feeds up from the past and into our present human culture. The products of former people are nutrients to us.

Fairly recently, soil ecologists have revealed a new level of connection among many types of trees. The intimate association of trees via their roots with the so-called mycorrhizal fungi has been long known. But the dynamics of the association are difficult to study because the fungi are so tiny—a fraction of a human hair in diameter—and therefore are nearly invisible thin threads that radiate in huge numbers outward into the soil from their more visible clumps, upon (and even within) the roots of nearly all species of trees.

The soil ecologists labeled with radiocarbon some of the carbon dioxide taken in by a tree so that they could track its path as it became carbon compounds traveling within the tree's aboveground structures (in the phloem), then down through its roots, and finally even into the threads of the attached fungi. Such transfer of food from tree to fungi had been known. It's the price the tree is willing to pay for the benefit from the fungi of increased nutrient uptake of mineral elements from the soil, which the tree also needs. The surprise came when some of the labeled carbon appeared meters away in the soil and even in neighboring trees. The fungal threads must have been acting like highways for the carbon, transporting it as far as trees of different species. Terms like *original world wide web* and *wood wide web* have made their way into the news and scientific literature. Obviously the separate trees were much more intimately linked than anyone had suspected.

I like to think of these belowground connections to the aboveground trees as analogous to human society. In the abbey the image of the extended self is intensified by the overwhelming presence of the influences represented by those buried there and the many human endeavors they pursued. We all share nutrients, both with one another now and from others of the past to others now. We all intake the dead and then pass them among us.

Recycling of the dead through the world's biogeochemical

cycles increases the overall, global abundance of life by about 200 times, as we saw in chapter 4. How much amplified is our degree of consciousness because we are social beings who live with others, learn from the past, and pass ideas around? Is it 200 times more consciousness? Many times that? I would hesitate to put a number on it, but the amplification is obviously great. A child deprived of social contact becomes a wild child, without language and with limited empathy for others. How lucky we are to be living among other people with whom we pass ideas into the psychological nutrient pool of sound, words, and print; of support, love, and even challenges; to ingest them as we need for growth. The soil, atmosphere, and ocean is an integrated system filled with life, and it serves as a metaphor for just how much more lively, how much more conscious, we are as a social system as a result of the cycling of the symbols, emotions, and values passed around through the globe-circling loops of our communication currents. We are richer because of the circuits of connections, both face-to-face and more distant across both space and time.

While we are alive we also comprise parts of memories in others, as well as the memory of others being in us. After we die we continue solely in others, a severe diminishment of our condition. Of course, the key is how we affect others, not just their memories of us. But we have these effects on others while we live as well, and after we die our influence fades and fades. We do return to lesser units. That's reality. But at least now, when alive, we can concentrate on the psychological circuits.

We can lament the myxobacterium in the support stem of the fruiting body, the Pacific salmon rushing upstream, the bristlecone pine tree near the end of a several-thousand-year lifetime, the enucleated red blood cells in our body. Yet we know that other bacteria, salmon, trees, and blood cells will rise again, similar to those that died. This renewal depends on the cycles of matter, the dead and the wastes turned back into life by life. So, too, we can learn to see our psychological selves. Ecology can serve as a metaphor for our thoughts, for our unconscious selves that have memories bound into ideas, emotions, and shared activities with others. Bacterium, salmon, tree, and blood cell—these live connected in webs of amplification for all life. They die. But while alive they thrive in the interconnected dignity of the circulating lives and deaths of others. When they die they stay in the web. They were each unique vibrant entities. When they die they retain a kind of living presence because of their connection to the ongoing whole.

So, too, with those in the psychological web of amplification of consciousness. Each of us contributes. And each of us should be thankful for all those in the past, ever newly circulating and forming our world. Each cannot be a Darwin in science, a Beethoven in music, or a Buddha in religion. To use an analogy from the ocean, most of us are not whales, but plankton. Yet each plankton cell is an integrity of being. It both takes from the whole and gives to the whole.

As a material pattern I am not truly immortal like the hydrogen nuclei in the water I drink, presences across nearly 14,000 million years. But my chemicals will circulate in the biosphere and become clouds and oceans and many wondrous creatures. I am also a cultural knot in a vast network. I am a cultural pattern. I have influenced others, others have influenced me. I see myself as a ripple on flowing water. I am here but not here. I am here merged into others more completely and subtly than language allows us to express.

We exist in a sea of mind; physical brains are molded by their development within the sea, creating knots of experience that are tied together. When one brain blinks out, so unravels the complex knot of mind born from the workings of that particular brain. But the larger sea and its back-and-forth echoes of causation between the knots continue. To fulfill our desires to live on after death, we need only probe more deeply into what we actually are as individual selves. We will find that we are not encased in skulls. We are extended out—indeed substantially merged with others. When I die, patterns of myself—say, styles of thinking about science—will continue in some of the younger scientists who studied with me (as I, before them, gained from my mentors). And so on, and so on. I am not bound to a single brain. I am right here in this book, and bits of me will come to life afresh in your own mind, should you offer me welcome.

The sea is one possibility for a metaphor of the extended self. But I like the river even better, because it is flowing, like history. Look now, in your mind's eye, at a gently moving stream or river. Its surface is not flat, but sloped, of course, in the direction of travel, a travel made possible by the collected body of water. On the water's surface are patterns. These are not exactly like tiny versions of the ocean's waves upon a beach, which are relatively distinct as they crest and crash, one by one. Rather, on a river the undulating, subtle hills and valleys on the water's surface are ripples. They seem to be distinct things because we call them by a name: ripples. Yet are they distinguishable from one another? Try. River ripples escape definition by being so frustratingly difficult to describe or visually freeze, especially the closer one looks. The ripples are present as individuals, and yet not there, but only interwoven patterns of the whole. They merge and diverge in a complex whole, the perceiving of which can alter one's state of consciousness.

The seventeenth-century Japanese haiku master Bashō, wrote some of the world's greatest poetry, often revelations about the Zen-like intensity of his moments in nature. Just days from death, Bashō was ill and in great pain. He wrote this poem:

Autumn deepens
My neighbor
What does he do?

Bashō's autumn referred to his final season of life. Is he driven inward by terror? Not at all. He asks about his neighbor. This is the extended self in highly conscious action. I am my neighbor, my community. I say hello as I say good-bye. To me, Bashō shows it is possible to seek a strong sense of community along with a healthy sense of individual accomplishment in the face of ultimate nothingness. Until then, let us connect— even with death in all its great journeys across time in its many evolving faces, for these can take us face to face with all of life.

REFERENCES

I have divided the references below into two categories for each chapter: (1) Names and Quotes, to provide sources for quoted material and for names whose work or ideas were specifically noted in the text, and (2) Further Reading, to point out a few readings that will augment the text. Additional explanation or elaboration of the subjects discussed can easily be found through Web searches. My aim has been to provide the big-picture, systems view of death without every tempting detour or possible nuance. In the end, all errors and other problems or infelicities are, of course, my responsibility.

OPENING EPIGRAMS

Buddha: in Karen Armstrong, *Buddha* (New York: Viking Penguin, 2001), 187. Hakuin: in Hebiichigo, translated by Philip B. Yampolsky, *The Zen Master Hakuin: Selected Writings* (New York: Columbia University Press, 1971), 219. Carl Sagan: *Billions and Billions: Thoughts on Life and Death at the Brink of the Millennium* (New York: Random House, 1997), 214.

1. IMPERMANENCE

Further Reading. For a classic account of carbon monoxide poisoning, check out the book *Alone* (Washington, DC: Island Press/Shearwater Books, 2003 [1938]) by Richard E. Byrd, an early Antarctic explorer who shared a winter isolated in an underground habitat with what he discovered was a malfunctioning stove. It's a gripping read.

2. EPICUREAN DANCE OF FRIENDS

Names and Quotes. Thomas Jefferson, in a 1819 letter to William Short, wrote, ". . . I too am an Epicurean." Epicurus: see his Letter to Menoeceus, the so-called Vatican Sayings (Number 52), and Principal Doctrines (Number 27). Beware: You might get interested in nuances of translation and then in Epicurus in general.

Further Reading. Before he died, a book John Richards took delight in editing was published: Arthur Versluis, Lee Irwin, John Richards, and Melinda Weinstein (editors), *Esotericism, Art, and Imagination* (East Lansing: Michigan State University Press, 2008). For a history of religious doubters across all the world's traditions, including the role of Epicurus and other early Greeks and Romans, see the superb book by Jennifer Michael Hecht, *Doubt: The Great Doubters and Their Legacy of Innovation from Socrates and Jesus to Thomas Jefferson and Emily Dickinson* (New York: HarperOne, 2003).

3. ORIGIN OF LIFE AS ORIGIN OF DEATH
Further Reading. Bacteria numbers: William B. Whitman, D. C. Coleman, and W. J. Wiebe, "Prokaryotes: The Unseen Majority," *Proceedings of the National Academy of Sciences, USA* 95 (1998), 6578–6593. Early viruses: Roland Wolkowitz, molecular virologist at San Diego State University, said, "I believe viruses have been here forever." *New York Times*, Tuesday, May 5, 2009 (Science Times section). Carl Zimmer, in his book *Microcosm: E. Coli and the New Science of Life* (New York: Pantheon, 2008), explains the amazing discoveries that have come from that single type of bacterium.

4. RECYCLING OF THE DEAD
Further Reading (and video viewing). For more on one of the world's wonders, the global carbon cycle, you might try my book CO_2 *Rising: The World's Greatest Environmental Challenge* (Cambridge, MA: MIT Press, 2008), and the fine book by Eric Roston, *The Carbon Age: How Life's Core Element Has Become Civilization's Greatest Threat* (New York: Walker & Company, paperback 2009). I have also made YouTube videos on the carbon cycle (search "Tyler Volk" in YouTube).

5. SUICIDAL BACTERIA
Names and Quotes. Carl Sagan, in *Cosmos* (New York, Ballantine: 1985), 20.
Further Reading. For an update on the scientific quest to understand how and why free-living plankton commit programmed suicide, see Nick Lane, "Origins of Death," *Nature* 453 (2008), 583–585.

6. LITTLE DEATHS IN BIG BODIES
Names and Quotes. The Robert Frost poem is "The Road Not Taken." Information "according to Margulis" was via personal communication. Guido Kroemer, "Mitochondrial Implication in Apoptosis: Towards an Endosymbiotic Hypothesis of Apoptosis Evolution," *Cell Death and Differentiation* 4 (1997), 443–456. Edmund K. LeGrand, "An Adaptationist View of Apoptosis," *Quarterly Review of Biology* 72 (1997), 135–147.
Further Reading. A Web search will reveal fascinating new discoveries on the regeneration or turnover of nerve cells, and now heart cells (approximately 0.5 percent per year). An interest in cell suicide in animal cells will quickly lead to the term *apoptosis*. For more on mitochondria, see Nick Lane's *Power, Sex, Suicide: Mitochondria and*

the Meaning of Life (New York: Oxford University Press, 2005). Any of Lynn Margulis's books will give you much food for thought about symbiogenesis; for example, *The Symbiotic Planet: A New Look at Evolution* (Chicago: Trafalgar Square, 1998), or any of her books coauthored with Dorion Sagan.

7. Built from Death

Further Reading. For studies on the zinnia: Hiroo Fukuda, "Programmed Cell Death During Vascular System Formation," *Cell Death and Differentiation* 4 (1997), 684–688. For the current status of similarities and differences between plant and animal cell suicides, see Brett Williams and Marty Dickman, "Plant Programmed Cell Death: Can't Live with It; Can't Live Without It," *Molecular Plant Pathology* 9 (2008), 531–544.

8. Extreme Senescence

Names and Quotes. The "disposal soma" theory: Thomas Kirkwood and Steven Austad, "Why Do We Age?" *Nature* 408 (2000), 233–238.

Further Reading. For overviews on the theory of aging, I suggest: R. E. Ricklefs, "The Evolution of Senescence from a Comparative Perspective," *Functional Ecology* 22 (2008), 379–392; and Martin Ackermann et al., "On the Evolutionary Origin of Aging," *Aging Cell* 6 (2007), 235–244 (available free online at www3.interscience.wiley.com/cgi-bin/fulltext/118523892/PDFSTART). Lynn Margulis points out Donald I. Williamson's evidence that many animals with a larval stage of development might have evolved through a genome merger. See his "The Origin of Larvae," *American Scientist* 95 (2007), 509–517.

9. Tuning Longevity

Names and Quotes. The phenomenon of "negligible senescence": See Caleb E. Finch, *Longevity, Senescence, and the Genome* (Chicago: University of Chicago Press, 1990). For "Fly now, die later," see Donna J. Holmes and Steven N. Austad, "Fly Now, Die Later: Life-History Correlates of Gliding and Flying in Mammals," *Journal of Mammology* 75 (1994), 224–226. Steven N. Austad and Kathleen E. Fischer, "Mammalian Aging, Metabolism, and Ecology: Evidence from Bats and Marsupials," *Journal of Gerontology: Biological Sciences* 46 (1991), B47–B53.

Further Reading. Thomas Sherratt and David Wilkinson's *Big Questions in Ecology and Evolution* (New York: Oxford University Press, 2009) includes important chapters on both sex and aging. Work on negligible senescence continues apace; perhaps the naked mole rat will help us: Rochelle Buffenstein, "Negligible Senescence in the Longest Living Rodents, the Naked Mole Rat: Insights from a

Successfully Aging Species," *Journal of Comparative Physiology B* 178 (2008), 439–445.

10. DEATH'S CULTURAL BLOOMING

Names and Quotes. Nigel Barley, *Grave Matters: A Lively History of Death Around the World* (New York: Henry Holt, 1997), 54–55. Barbara J. King, *Evolving God: A Provocative View of the Origins of Religion* (New York: Doubleday, 2007); I also drew from King's insightful book for the story about the chimp death. Steven Mithen, *The Prehistory of the Mind: The Cognitive Origins of Art, Religion, and Science* (London: Thames and Hudson, 1999 [1996]).

Further Reading. The crosscultural comparisons of ritual sacrifices of animals and humans is a fascinating topic for another day or book. See, for example, Walter Burkert, *Homo Necans: The Anthropology of Ancient Greek Sacrificial Ritual and Myth* (Berkeley: University of California Press, 1986), or the "book" of the Bible called Leviticus.

11. DEATH-DENYING DEFENSES

Names and Quotes. The phrase *worm at the core* was used by William James, *Varieties of Religious Experience: A Study in Human Nature* (New York: Mentor Edition, 1902 [1958]), 121; personal communication from Jeff Greenberg. For a general audience overview, see Sheldon Solomon, Jeff Greenberg, and Tom Pyszczynski, "Tales from the Crypt: On the Role of Death in Life," *Zygon Journal of Religion and Science* 33 (1998), 9–43.

Further Reading. Thomas Pyszczynski, Sheldon Solomon, and Jeff Greenberg, *In the Wake of 9/11: The Psychology of Terror* (Washington, DC: American Psychological Association, 2002). This book provides a detailed overview of the many experiments in terror management theory, with applications to the problems of political and religious terrorism. The authors are at work on a forthcoming book for general readers, to be published by Random House, tentatively titled *The Worm at the Core*.

12. DEATH AND SELF-ESTEEM

Names and Quotes. Sigmund Freud, in a letter to Marie Bonaparte, August 13, 1937: www.pep-web.org/document.php?id=zbk.051.0436a (accessed June 8, 2009). Experiments with mirrors: Jamie Arndt, et al., "Terror Management and Self-Awareness: Evidence That Mortality Salience Provokes Avoidance of the Self-Focused State," *Personality and Social Psychology Bulletin* 24 (1998), 1216–1227.

Further Reading. Jeff Greenberg, Spee Kosloff, Sheldon Solomon, Florette Cohen, and Mark Landau, "Toward Understanding the Fame Game: The Effect of Mortality Salience on the Appeal of Fame,"

Self and Identity, in press; to the authors' knowledge this paper is the first to experimentally investigate what determines interest in fame and celebrity, and in a series of experiments they do find that death awareness increases a variety of measures of attraction to fame and celebrity. David Loy, in chapter 5 of *Lack and Transcendence: The Problem of Death and Life in Psychotherapy, Existentialism, and Buddhism* (Amherst, NY: Humanity Books, 1996), provides an analysis of fame and other psychological drives related to seeking forms of immortality as a protection against a general "lack" we feel in life, some of which is related to mortality.

13. RIPPLES UPON FLOWING WATER

Names and Quotes. Bashō poem: Robert Aitken, *A Zen Wave: Bashō's Haiku and Zen* (New York: Weatherhill, 1989 [1978]). Translator Robert Aitken's analysis confirms that Bashō's autumn referred to his life's seasonal completion. The phrase *wood wide web* was first used by T. Hegason, et al., "Ploughing up the Wood-Wide Web?," *Nature* 394 (1998), 431, to describe the research by S. W. Simard, et al., "Net Transfer of Carbon Between Ectomycorrhizal Tree Species in the Field," *Nature* 394 (1998), 579–582, which has been cited in the scientific literature more than 230 times.

Further Reading. Douglas R. Hofstadter puts forth his fascinating case that we are vastly interconnected beings wherein mind is not limited to the skin. See his book *I Am a Strange Loop* (New York: Basic, 2007).

ACKNOWLEDGMENTS

In *Death* I wanted to convey the many ways death is indispensable to life and has been over the course of evolutionary time since life's beginnings. Considering the many faces of death, as Dorion Sagan has done with sex in the other half of this volume, allows us to see how the Grim Reaper has changed in both its evolved and inadvertent facets across the great epic of time.

For this book I was able to draw upon, among other works, discussions in a book I wrote some years ago that was nearly three times longer, *What Is Death?: A Scientist Looks at the Cycle of Life* (John Wiley & Sons, 2002). Of that earlier work, it could be said what the seventeenth-century French mathematician and philosopher Blaise Pascal said of one of his letters (from *Provincial Letters XVI*): "I made this letter very long because I did not have the leisure to make it shorter." I suspect that many writers have felt similar sentiments. If only one had had additional time to make a piece more concise, and, presumably, more cogent, or at least more accessible, especially in today's hurried world. I am fortunate to have been given such an opportunity, and I have not only written a far shorter book but reconceived the narrative along evolutionary lines and included much new material.

For giving me this opportunity and for enthusiastic help along the way, I thank Margo Baldwin and Joni Praded at Chelsea Green Publishing Company. Dorion Sagan, author of the other side of this unique double book, has been a delight to work with, and I want to acknowledge Dorion for his "out-of-the-book" idea to put two short works together into one. Lynn Margulis, who with Dorion co-directs the Sciencewriters imprint at Chelsea Green, graciously went through the entire first draft, noting technical glitches and providing writerly advice; I am grateful to her inspiration in other ways, too.

Amelia Amon read through a number of chapters and was always there with curiosity and dependable support. Peter Nevraumont facilitated the use of my previous book, and provided much encouragement. Geologist Michael Rampino checked my discussion in chapter 3. I thank Marty Hoffert, Lyn Hughes, Joseph LeDoux, Susan Doll, John Horgan, Jill Neimark, Jeff Bloom, my sister Kristin Volk Funk, and all other dear friends and family for ongoing discussions about the meaning of life. Jeff Greenberg of the University of Arizona, one of the researchers in terror management theory, has been an invaluable friend in deepening my understanding of how the knowledge of mortality affects our day-to-day lives. He and the other researchers are amazing scientists and fun individuals of the highest order. Finally, I note with profound gratitude the developmental editing of Jonathan Cobb, who truly cares.

—Tyler Volk

INDEX

ABOUT THE AUTHOR

Tyler Volk is Science Director for Environmental Studies and Professor of Biology at New York University. Recipient of the NYU All-University Distinguished Teaching Award, Volk lectures and travels widely, communicates his ideas in a variety of media, plays lead guitar for the all-scientist rock band The Amygdaloids, and is an avid outdoorsman. Volk's previous books include *CO$_2$ Rising: The World's Greatest Environmental Challenge*; *Metapatterns Across Space, Time, and Mind*; and *Gaia's Body: Toward a Physiology of Earth*.

AMELIA AMON

green
press
INITIATIVE

Chelsea Green Publishing is committed to preserving ancient forests and natural resources. We elected to print this title on 30-percent postconsumer recycled paper, processed chlorine-free. As a result, for this printing, we have saved:

12 Trees (40' tall and 6-8" diameter)
5,297 Gallons of Wastewater
4 Million BTUs Total Energy
322 Pounds of Solid Waste
1,100 Pounds of Greenhouse Gases

Chelsea Green Publishing made this paper choice because we and our printer, Thomson-Shore, Inc., are members of the Green Press Initiative, a nonprofit program dedicated to supporting authors, publishers, and suppliers in their efforts to reduce their use of fiber obtained from endangered forests. For more information, visit: www.greenpressinitiative.org.

Environmental impact estimates were made using the Environmental Defense Paper Calculator. For more information visit: www.papercalculator.org.

Two books in one, *Death & Sex* explores the two facts of life that have dominated our thoughts, fears, and dreams since the earliest days of human history.

In *Sex*, Dorion Sagan takes a delightful, irreverent, and informative romp through the science, philosophy, and literature of humanity's most obsessive subject. Linking evolutionary biology to salacious readings of the lives and thoughts of such notables as the Marquis de Sade and Simone de Beauvoir, *Sex* touches on a miscellany of interrelated topics ranging from animal genitalia to sperm competition, from the difference between nakedness and nudity to the origins of language, from ovulation to love and loneliness.

Praise for *Sex* by Dorion Sagan

"What delicious writing and reading! I love this wise and funny big-little book."   —ERICA JONG, bestselling author of *Fear of Flying* and *Seducing the Demon*

"In *Death & Sex* two of my favorite thinkers and writers ruminate on two of my favorite subjects and turn up all manner of unexpected interconnections. The result is a splendidly entertaining, informative, and original piece of science writing."   —JOHN HORGAN, author of *The End of Science* and *Rational Mysticism*

"I genuinely can't recall reading a more inspiring or entertaining book in years!"   —FRANK RYAN, author of *Virolution* and *The Forgotten Plague*

"In just 100 pages, everything you really need to know about sex: Why? When? Where? With whom? Dorion Sagan slides effortlessly from seductive prose to bringing the reader sharp up against one astonishing scientific discovery after another."
—DENIS NOBLE, Fellow Royal Society and Professor Emeritus, Oxford University

"In *Sex*, Dorion Sagan writes with a wit that no other science writer of our generation can equal."   —HOWARD BLOOM, author of *The Lucifer Principle: A Scientific Expedition into the Forces of History*

"Dorion Sagan muses ruthlessly on the topic of sex and the result is as twisted and tangled as a set of bedsheets. Hyena sex,

cycad sex, lots of primate sex, and even a digression on why the Marquis de Sade was not such a bad guy—Sagan takes pleasure in revealing it all. He even makes bacterial sex sound fun."
—BETSEY DYER, Professor of Biology, Wheaton College, author of *A Field Guide to Bacteria*

"Sagan shows us just how deep the riddle of sex goes—pulsing through the world from the Marquis de Sade's plays right down to the bacteria that make up our cells. This slim book allows us to be voyeurs and exhibitionists. . . . Whether you end up resonating more readily with the puritanical tendencies of the orangutans or with the orgiastic culture of the bonobo chimps, Dorion Sagan's *Sex* will provide a hilarious, thoughtful, and unforgettable time. It's more fun than my day job."—CONNER HABIB, porn star and writer

"In *Death & Sex*—two books in one—quotidian simplicities are dissolved in the acid of evolutionary theory. Death turns out to be more complicated than to be or not to be; and sex is seen to be far more complicated than a tale about a man, a woman, and a garden snake. Together, they form a pair of insightful lessons in the application of Darwinian concepts."   —ANDREW LIONEL BLAIS, author of *On the Plurality of Actual Worlds*

"A boisterous Siamese twin of a book which looks at the two sides of the same molecular process—that of sex and that of death—within the framework of life almost eternal. Enjoy, and know you are part of it."   —CRISPIN TICKELL, Director of the Policy Foresight Programme, Oxford University, and former UK Ambassador to the United Nations

"Dorion Sagan and Tyler Volk show us sex is optional and death is necessary, turning the tables on our lusts and fears, our origins and endings, in a surprisingly enticing way."
—ADAM DANIEL STULBERG, *Poetic Interconnections*

"Eschewing taboos and transgressing disciplinary boundaries, this volume manages to be, at once, both playfully iconoclastic and technically informative. Where else is one going to experience such from chance encounters with de Sade, Monty Python, Bashō, and Poincaré?"   —SIMON GLYNN, Professor of Philosophy, Florida Atlantic University

Sex

Sex

DORION SAGAN

 A Sciencewriters Book

CHELSEA GREEN PUBLISHING COMPANY
WHITE RIVER JUNCTION, VERMONT

A Sciencewriters Book

scientific knowledge through enchantment
Sciencewriters Books is an imprint of Chelsea Green Publishing. Founded
and codirected by Lynn Margulis and Dorion Sagan, Sciencewriters is an
educational partnership devoted to advancing science through enchantment in
the form of the finest possible books, videos, and other media.

Project Manager: Emily Foote
Developmental Editor: Jonathan Cobb
Copy Editor: Laura Jorstad
Proofreader: Helen Walden
Indexer: Christy Stroud
Designer: Peter Holm, Sterling Hill Productions
Cover Design: Kelly Blair

Printed in the United States of America
First printing September, 2009
10 9 8 7 6 5 4 3 2 1   09 10 11 12 13

Our Commitment to Green Publishing
Chelsea Green sees publishing as a tool for cultural change and ecological stewardship.
We strive to align our book manufacturing practices with our editorial mission and to
reduce the impact of our business enterprise in the environment. We print our books and
catalogs on chlorine-free recycled paper, using vegetable-based inks whenever possible.
This book may cost slightly more because we use recycled paper, and we hope you'll
agree that it's worth it. Chelsea Green is a member of the Green Press Initiative (www.
greenpressinitiative.org), a nonprofit coalition of publishers, manufacturers, and authors
working to protect the world's endangered forests and conserve natural resources. *Death
& Sex* was printed on Natures Natural, a 30-percent postconsumer recycled paper supplied
by Thomson-Shore.

Library of Congress Cataloging-in-Publication Data
Volk, Tyler.
  Death & sex / Tyler Volk and Dorion Sagan.
    p. cm.
  Includes bibliographical references and index.
  ISBN 978-1-60358-143-1
1. Death. 2. Aging. 3. Sex. I. Sagan, Dorion, 1959- II. Title. III. Title: Death and sex.

  QP87.V647 2009
  612.6'7--dc22

                                    2009030141

Chelsea Green Publishing Company
Post Office Box 428
White River Junction, VT 05001
(802) 295-6300
www.chelseagreen.com

LORD DARLINGTON: I couldn't help it.
I can resist everything except temptation.

—Oscar Wilde, *Lady Windermere's Fan*

I am as pure as the driven slush.

—Tallulah Bankhead (1902–1968)

Nothing in the world is single;
All things by a law divine
In another's being mingle—
Why not I with thine?

See, the mountains kiss high heaven,
And the waves clasp one another;
No sister flower could be forgiven
If it disdained its brother;
And the sunlight clasps the earth,
And the moonbeams kiss the sea;—
What are all these kissings worth,
If thou kiss not me?

—Percy Bysshe Shelley,
"Love's Philosophy"

If all the young ladies who attended the Yale prom were
laid end to end, no one would be the least surprised.

—Dorothy Parker

How is it that, in the human body, reproduction is the
only function to be performed by an organ of which
an individual carries only one half so that he has to
spend an enormous amount of time and energy to
find another half?

—François Jacob, *The Possible and the Actual*

CONTENTS

A BRIEF HISTORY OF THE DIRTY DEED*

*J*ust yesterday I came out of a coffee shop, walking between a gaggle of giggling girls leaning over a balcony—really the railing of a long wheelchair ramp—and the two boys they were flirting with, seated on benches on the sidewalk below. As I walked between the girls on their impromptu balcony and the boys on the street, I noticed a blue condom in its package on the sidewalk and kicked it without breaking stride. Sure enough, that had been the source of the excitement. I heard more laughter as I kicked it again, less obviously an accident this time, and ignored calls to pick it up. As torn condoms and unplanned children attest, and despite the roadblocks and countermeasures, from contraception to warnings of the dire consequences in store for those who so much as play with themselves or engage in premarital relations, let alone more exotic practices, sex finds a way. It exerts its powerful pull. Gay, straight, bi-, or non-, we are all, even if only mentally, under the influence of this basic biological urge.

Other than making a living, and dying, few subjects wax as huge over the human psyche as sex. Questions of attraction, power, abuse, dating, self-image, family, and the vicarious immortality of having children and grandchildren all hinge upon the sexual relation. The teenager's body morphs like something out of a bad movie, the appearance of new patches of hair both welcome as a badge of burgeoning adulthood and frightening as a reminder of a putative evolutionary

---

* Also known as "horizontal refreshment," "the four-legged frolic," a "flesh session," "Irish whist," "dropping a load of baby batter," "making the beast with two backs" (Shakespeare), and "doing the wild thing," the sex act reliably spawns a long list of clever euphemisms.

past in which ancestors were as ugly as sin—certainly no one you would want to take home to Mother.

Each of us is different, and that is in part the result of our sexual nature, which ensures that each new being will carry a random hand of genetic cards assembled together from the separate deals of each parent. Although the beautiful models on television may give us unrealistically high expectations of idealized and symmetrical beauty in our fantasized partners, in reality we are glad that each of us is not an identical copy in a cookie-cutter mold off an evolutionary assembly line. They may annoy us no end, in ourselves and others, yet it's the little things, from freckles to the shape of a lip, from a way of speaking to the curve of a hip, or the uniqueness of *his* words or *her* thoughts, that mark the loved one as *our* loved one, as opposed to some random drone. I love *you,* we all say, and sometimes we mean it. Sex produces the differences that we recognize as the one we cherish, the special qualities of the beloved in the eyes of the lover. When we love someone who loves us we want to live; unfortunately, as crimes of passion sometimes brutally attest, the opposite is also true.

Sex incites fantasies and families, delight and despair, despondency and ecstasy, swears and smiles. It is also a whetstone upon which writers have sharpened to a razor's edge the rapiers of their wit. Take Lord Chesterfield who, in the eighteenth century, in his *Letters to His Son on the Art of Becoming a Man of the World and a Gentleman,* cut to the chase in a sentence still not bested: "The expense is damnable, the pleasure momentary and the position ludicrous."

Chesterfield's droll observation highlights some deep truths about our status as living, breeding beings on this planet. The damnable expense—which in Chesterfield's case doubtless refers to the money and time spent in wooing, dating, and engaging in matrimony—applies to all sexually reproductive

organisms. Considering that some organisms can clone themselves—a well-fed ameba simply grows and splits to produce two new amebas—what is the point of mating and dating, finding and grinding? Why all this aggravation to court and couple?

When evolution can take one organism and create two, why make matters more difficult by requiring two organisms to make one?

When we ask, "Why sex?" the answer seems obvious: "To reproduce."

But the curious mind presses on: "Why reproduce?"

In fact there may be a method to nature's madness, her wanton expenditure of energy and effort: advantages beyond the obvious fact that there is (as yet) no other viable way to make babies.

Sex and reproduction are not necessarily connected, even though they are strongly linked in our species as well as in most plants and animals. In biological terms, *sexual* reproduction can be defined as "the formation of new individuals from the genes of at least two different sources"—for you, your biological parents. *Simple* reproduction, by contrast, is "an increase in the number of individuals"—but those individuals don't necessarily have to pick up new genes.

Like most things on this planet, sex started long before we did. Bacteria have been exchanging genes, without needing one another to reproduce, since billions of years before the evolution of plants and animals. If the methane recently reported from Mars turns out to have been produced by bacteria, it could even be that life evolved elsewhere in deep space, and that the methane-producing bacteria, considered among the most anciently evolved life-forms, were even sexually trading genes before the origin of the solar system. Sex, in other words, although of the bacterial kind, might be older than life on Earth! In any case, bacteria were getting it on, trading

DNA, for maybe two thousand million years before the earliest ameba-like cells appeared. It is in these cells—more complex than bacteria--that our kind of sex, called meiotic sex, which usually requires two parents, first evolved.

But swinging bacteria started the party. Genes transfer among them without their having to reproduce. In some cases, one "parent" in an act of bacterial sex is not even alive; it's simply a raw gene—a DNA molecule in solution. This phenomenon, called the transforming principle, was first demonstrated by British medical officer Frederick Griffith in 1928. Griffith found that even dead bacteria of one strain could pass on their genetic material to live bacteria of another strain, thus transforming their offspring into the strain of the dead bacteria. It was later discovered that the transformation was actually caused by the living bacteria absorbing the DNA of the dead bacteria and using it to replicate. Such necrophilia in reverse, like a corpse impregnating a coroner, is typical of the diverse possibilities of sex in the natural world.

We now know that viral DNA, genetic elements called plasmids, and whole bacteria with an entire set of genes may also serve as "parents" in bacterial sex. This kind of sex is infective, direct; it doesn't have to wait around for reproduction to transfer genes. (Scientists call it horizontal gene transfer, because the genes move directly from cell to cell—as opposed to vertical transfer, where the genes are handed down only from parent to offspring.) In bacterial sex, an old being becomes new without an increase in the number of individuals: It is sex *without* reproduction. The results of bacterial sex are bacteria with some new genes. Then, when these sexed bacteria reproduce, they pass to their offspring the new abilities or traits conferred by their new genes.

If we had sex like this, a brunette with brown eyes could settle into a Jacuzzi with a blue-eyed redhead and leave with

red hair and blue eyes; what's more, she could pass the new traits on to her children.

Bacterial sex, although it is not needed for reproduction, may play a major role in evolution. Infective viruses, which can pass on bacterial genes, are not always harmful. They sometimes spread useful genes, including resistance to other viruses. Also, by trading genes, bacteria can adapt to environmental change far more rapidly than larger organisms, which must reproduce first. Bacteriologists have compared the worldwide capacity of bacteria for genetic exchange to a planetary intelligence, and to the Internet.

Tantalizing evidence from studies of bacteria exposed to ultraviolet rays suggests that bacterial sex (again, not our sort of two-parent, reproductive sex) evolved in the two billion years before Earth was enveloped in a protective ozone layer. This thin layer of O_3 is thought to have only appeared one to two billion years ago as a by-product of the metabolic process of turquoise-colored bacteria—cyanobacteria—whose energy comes from sunlight. The ozone layer about the Earth today shields life from the vast majority of damaging ultraviolet radiation. Bombard modern bacteria with levels of ultraviolet radiation similar to those before the ozone layer formed, however, and the bacteria disperse bits of naked and protein-coated DNA and RNA into the surrounding medium. In the old, pre-ozone-layer days, more sunlight would have penetrated to spur bacteria to eject their genes, leading to recombination orgies in broad daylight.

The sex that humans and other animals, as well as plants and fungi and some ameba-like organisms, engage in is decidedly different. Bacteria have no nuclei (the dense centers of cells) and no true chromosomes (DNA packaged in protein strands), whereas beings like us, made of larger cells, do. These larger cells, containing nuclei and other complex structures enclosed

in membranes, are called eukaryotic (from the Greek for "nut," referring to the nucleus). Beings composed of such cells range from single-celled *Paramecium* to Britney Spears, made of trillions of eukaryotic cells. Whereas bacteria can take in anything from a single gene floating loose in the surrounding medium to all of another individual bacterium's genes, the reproductive sex of cells with nuclei typical of nonbacterial species, including ours, involves the doubling of chromosomes and, usually, the merging of nuclei, one from each of two parents, into one. There are no deadbeat dads among bacteria because they all are single parents, and none uses sex to make a "baby." In two-parent beings like us, however, the sexually merged nucleus with two sets of chromosomes must get back to its original state (that of the sperm and egg cells) with only one set of chromosomes.

In evolutionary terms, you are a sperm's or egg's way of making another sperm or egg, just as a hen is an egg's way of making another egg or a butterfly is a butterfly's genes' way of making more butterfly genes. Sex is, so far, an inescapable part of this cycle. The multiplication of cells in a growing animal reaches a peak and levels off. For more animal lives to continue, the growth cycle must be reset. Sex does the trick. Behind the romance and tears, the love and fears, then, is the romantic irony of a transitory consciousness that can be cavalierly discarded after completion of the reproductive cycle.

The eggs and sperm each have one set of chromosomes, and it is their need to get together, once every generation, that causes boy and girl trouble. While the dying body ultimately perishes, the species form, maintained by erotic urges, continues. Of course, people may have sex for all sorts of reasons—to show love, for pleasure and recreation, to feel power, to bestow themselves as a gift, for exercise, even to get rid of headaches—other than reproduction. Neither condoms nor same-sex dalliances increase the number of people: to the

contrary. But for hundreds of millions of years animal evolution has been linked to sex. To understand it, we must look into what Shakespeare called the "backward abysm" of time. In truth, to understand our sex we must look at other organisms, from the apes whose behaviors illuminate our own to the ameba-like cells in which two-parent sex evolved. Only then can we understand the damnable expense, the fleeting pleasure, and the ludicrous position.

Humans and Other Chimps

Forbidden Fruit

*W*hen I was newly married, driving in Florida after a
sparsely attended shotgun wedding (just the two of
us and a justice of the peace), a preacher came on the radio.
I listened because he was criticizing efforts to understand the
evolution of sexuality while I was engaged, as a junior science
writer, in writing a book on it (with, of all people, my mother,
an evolutionary biologist). Scientists these days are taking it
down to ridiculous levels, beyond the level of the flea, he said
with scorn in his voice. By God, they were even trying to look
to bacteria for answers! Listen, he continued. You don't need to
look at the birds and bees, let alone microorganisms, to understand sex. Everything you need to know about the subject is
already there, written for you in black and white, in the Bible.

As a northerner in the Bible Belt I was perturbed. *The
Origins of Sex: Four Billion Years of Genetic Recombination* had
yet to come out. Highly technical, due to be published by
Yale University Press, this book, just as the Christian broad-
caster warned, took it down to the level of cells. Who was I, a
twenty-six-year-old, to have such hubris?

Although I'd never read the Good Book cover-to-cover (I
hear there are some bawdy parts), and had been brought up
by scientists (astronomer father, chemist stepfather, and biologist mother), I could not help but feel accused by this stranger's sermon. In Genesis, as I understood it, Adam is made by
God in his image, Eve is taken from Adam's rib, and they live
happily ever after—at least until the Fall. As Jimmy Buffett

sings (which you can also hear driving through Florida), some say a woman is to blame: The fall is Eve's fault, as it is she who let the trickster snake whisper sweet somethings in her ear and yielded to the temptation to munch of the sumptuous fruit of the Tree of Knowledge of Good and Evil upon whose branches he hung. She took her fateful bite, and the rest is history.

Now, in English texts such as the King James translation of the Bible, the fruit she bit is an apple, but some say apricots, pomegranates, figs, or grapes were more likely the fruit of the one tree God prohibited the first couple from eating in Genesis 2:9. According to ethnobotanist R. Gordon Wasson, the "apple" may even have been a white-spotted red mushroom, *Amanita muscaria,* of the sort that the hookah-smoking snail sits upon in *Alice's Adventures in Wonderland.* Forming a symbiotic partnership with the roots of trees, this fungus is a kind of "fruit." It also qualifies as a candidate for the first bite on the grounds of being psychoactive and poisonous, although for sheer salacious lubriciousness in cross section it's hard to top the apple.

Who could blame Eve, surrounded by all those arrogant males, for taking a bite of the forbidden fruit? Even if the main thing learned from that luscious bit of nutriment was the revelation that they were naked. Nonetheless, for her contagious disobedience in partaking of such a licentious snack, God the Father doled out a suitably agricultural punishment: They were to toil with the soil, and grow their own, rather than continuing on as freeloaders in a paradise they didn't appreciate, blithely violating divine edicts, like the prohibition against education.

According to the Bible, this was the female-precipitated ur-disaster for which we continue to pay. There was also said to be a Tree of Life in the Garden that conferred immortality, but God made sure that Adam and Eve, given their sinful natures,

didn't get a piece of that. Instead, they were expelled from the Garden, fell to Earth (or, more allegorically, into incarnation and time), and were subject henceforth to aging and death.

Well, maybe. There does seem to be a connection, and not just in the Bible, between sex and death. The tiny ameboid microbes that preceded all animals have chromosomes with DNA in the nuclei of their cells. Such cells, bigger than bacteria, don't all mate, but some do. And when they do, parts of the cell of one, the oxygen-using mitochondria, must be "put to death" by the other. When an egg and sperm merge, like a young couple moving into a Manhattan apartment, they can't take everything with them. Some stuff, such as his DNA-containing mitochondria, never make it into the fertilized cell. Of the trillions of cells of our bodies, only a few sperm and eggs survive into the next generation. In coming together in reproductive sex, the sex cells leave male and female bodies behind to grow a fresh being. It is the reproductive cycle, not the individual animal, that is selected for over evolutionary time. After the midair mating of a queen by a horny honeybee, the latter goes *pop,* audibly, as its penis breaks off inside her (blocking passage to other would-be suitors) while the rest of his body plunges to its death. It may seem tragic to have life cut short in such flagrant fashion. But then the honeybee exploding immediately prior to death is lucky relative to his fellows, who can number up to twenty-five thousand, all virgins whose efforts to compete for the queen's sexual favors fail, their entire lives an exercise in frustration.

Evolution travels light. Sex and death do go together, although the colorful stories of Genesis, written more than two thousand years ago, favor the story of a talking serpent over the fact of serpentine DNA, whose structure was deduced only in March 1953. Scientific stories about sex are not necessarily as pretty as Scarlett Johansson, as romantic as a honeymoon on O'ahu, or as memorable as Adam's de-ribbing. But

exploring the evolutionary story of our sexual nature based on science will help us get to the bottom of this topic better than the radio sermonizer's version of religion.

If we are to be punished for Eve's congress with the twisting reptile of the Tree of Knowledge, we should at least relish each morsel of wisdom that her sinking her incisors into the ripened red ovary of the flowering *Malus domestica*—the fruit of the apple tree—has made possible for us.

The Naked Truth

*I*t is the proverbial problem of the onion: Whenever one layer is stripped away, a new layer confronts us, until there is nothing left but the pursuit itself. So it is with the desire for knowledge, carnal and the regular kind: Getting there is not only half, it's all, of the fun.

Moved on the occasion of the death of Bettie Page, a 1950s movie star and pinup girl, Manohla Dargis (quoting art critic John Berger) reflected on the difference between nakedness and nudity: "To be naked is to be oneself. To be nude is to be seen naked by others and yet not recognized for oneself." Not being recognized means there is more in store.

Nakedness is pretty straightforward. It is exposure. What you see is what you get.

The allure of nudity is subtler. What is still concealed is equal to or greater than what is revealed. What you see is not what you get.

Partial revelation triggers temptation, seduction, the dance of desire. This is the principle of nudity. Even if what is revealed is the naked body, there is more than meets your eye. Nakedness, although seemingly so close to nudity, is in a way its opposite. Nakedness is our natural state, but it may show too much too soon, more than we want to see, destroying desire. The principle of desire is that people want what they can't have and, if they get what they want, they don't really want it. As long as you don't have what it is that you think you want, you can harbor the illusion that if you possessed it you

would finally be satisfied. Or, if in fact you really chose what you wanted, and through striving arrived at your goal to your complete satisfaction, you would likely turn your attention to new pursuits. Infinite desire, bored with completion, thrilling to the process, is fascinated with the partial revelation, the play of concealing and revealing described by the ever-tempting lure of the nude.

Nakedness is different. To understand it, we must go back to our furry ancestors and regard it in evolutionary terms. Animals, especially visually oriented animals such as ourselves, recognize one another by their surfaces. Color, pattern, and shape of skin and hair give a first impression, allowing an organism to peg another as belonging or not to the same species—a first step in identifying a possible mate. Sexual organisms judge one another, unconsciously calculating the possibilities of mixing lives and genes, by their appearances—not just how they look, but the sounds they make, how they smell, and what they do.

But because we wear clothes, we can modify our appearances in a manner unknown in the animal world. The changing surfaces of sartorial fashion have thrown a monkey wrench into ancient calculations of genetic worthiness. With clothes and cosmetics, appearances can be manipulated, the raw reality of our animal physiology tweaked and accentuated. Which is why we associate nakedness with truth: The naked is that which cannot be altered. It is where the paint stops.

Evolutionarily, the natural clothing of fur, or hair, appears to have been lost among our ancestors perhaps a quarter of a million years ago. Something made us naked, although pubic and underarm hair make a bit of a comeback at puberty.

Of all the estimated thirty million species on planet Earth, we have the greatest overlap of DNA with chimpanzees, *Pan troglodytes*. They are our closest relatives. The closest living relatives to human beings, as well as we ourselves, are grouped

into the hominoids ("man-like" beings). These include thirteen gibbon species and the siamang, collectively known as the lesser apes. Closer to us among the hominoids than these lesser apes are the great apes: the chimps, the gorillas, and the orangutans. The lifestyles, including the mating habits, of these wonderful but less populous primate brethren provide clues about the sex lives of the ancestral animals whose mate choices and subsequent frolicking helped shape *Homo sapiens*.

But what's likely to jump out immediately at us in comparing ourselves with the great apes isn't a similarity but a difference: We have almost no body hair next to them. Why is this? A first thought might be that hairlessness gives some kind of advantage to our species. But just because something exists in the animal world doesn't mean that it was selected for—that is, that it evolved because it conferred an advantage. This is an important principle to keep in mind when thinking about any aspect of our evolution, not just sex. Because more organisms in a species reproduce than can survive, the ones left tend to have features useful in survival. Over time inherited variation leads to evolutionary change. But some changes may occur not because they themselves have an advantage but because they are structurally connected to other features and simply change with those features. For example, there is no advantage to having red blood that we know of. It is simply the result of the electromagnetic wavelength reflected by hemoglobin, the iron-containing molecule in the blood. Some mammals might escape predators better if their blood were transparent. Why then does their blood remain red? We can invent a story, but there's probably no good reason. Rather, it's a structural limitation: The species in question don't have the option to switch to non-hemoglobin-based, un-red blood. Not every feature in the evolutionary change game is up for grabs.

Charles Darwin distinguished between natural selection and sexual selection. Natural selection proceeds through less

likely reproduction in a species of its less "adapted" organisms (organisms less able to survive in a given environment), leading to the survival of the rest. Sexual selection, resting on a competition among males for limited females and a selection among females for males with features they like, also leads to organic change. The concept of natural selection is based on the model of artificial selection, the breeding of pigeons and dogs with which Darwin was familiar. In natural selection, as its name implies, rather than conscious bird and dog breeders doing the selecting, nature does. For example, whereas dog breeders selected by drowning puppies that didn't match their expectations for specific physical traits, nature selected by tending to eliminate canines with an inferior sense of smell necessary to locate one another and prey, fleetness of foot needed to escape predators, and other factors. *Natural selection* is simply the name for the weeding out of organisms less able to survive and reproduce successfully in a given environment.

Sexual selection is similar, but arguably more conscious—as life-forms themselves exert choices that make a genetic difference. If peahens shop for males with iridescent peacock rainbows on their tail feathers, and spurn gray or insufficiently colorful males, their aesthetic choices may maintain and shape the look of their species. Aesthetic fashion in mating projects its offspring onto the stage of life.

Darwin suggested an important place for sexual selection in evolution. The peacock's plumage (partially mimicked by the cancan dancer or fop with a plume in his cap) can be explained by sexual selection: Peahens prefer brightly feathered peacocks. Females may choose males with splendid plumage in part because it advertises genetic robustness, as well as freedom from parasites, not to mention the fact that colorful peacocks, more obvious to predators, must compensate with relative strength and cleverness. Experimenters have actually taped additional tail feathers and colors onto male birds of various

species and measured their mating and reproductive success with females of the species. The birdbrains do not recognize that they have been tricked, but do preferentially mate with the made-over birds. So Darwin's basic idea has been experimentally confirmed.

Nonetheless, naturalist Alfred Wallace, Darwin's contemporary, argued that the part played by sexual ornaments paled in comparison with the role in sexual selection played by attributes useful in combat, agility, or bodily vigor in general. He doubted that insects, for example, could be sufficiently impressed by differences in color schemes or patterns to make selections based on them. All visible objects, such as chlorophyll-green grass and inanimate objects besides, had to have some color or another. Bright colors inside our bodies—for example, blue arteries and yellow bile—could not be selected. The colors of your pancreas will not give you an advantage, or even be noticed. For Wallace, body design in insects—as in bugs that look like sticks or leaves—was far more likely to be the result of natural selection preserving organisms that had camouflage than it was to reflect choices made by robotically dumb arthropod females.

So it is not enough to assume that, just because some odd or obvious feature exists in an organism, it was selected for. In some cases nature is stymied because she is working with limited materials. For example, having even bigger heads might have increased (due to increasing processing power of the brain) our carnivorous ancestors' ability to track down their prey, but the growth of ancestral brains was constrained by the size of the opening of the birth canal, the ability of those brains to grow outside the womb, and the ability of neck muscles to support big heads and durable skulls. So, too, males have nipples. Does that mean they were selected for because ancestral males without them were less able to survive? No. Males have nipples because females have nipples, which are

required for breast-feeding of infants. The fertilized egg that becomes the embryo is shaped by hormones, first in the mother's body and then in another wave at puberty. Hormones, like a resourceful artist, can do amazing things with limited materials. But they can't do everything. The nipples of the male are like the brushstrokes we see when looking close at a painting: They attest to basic materials, the foundation—the substance, not the selection.

Thinking along evolutionary lines about the survival value of independent organismic features can lead to surprising and fruitful explanations. But as we've just seen, it does not follow that every trait was actively selected for. Neither natural nor sexual selection can account for everything. The caution about attributing every feature in evolution to an active force of selection was a running theme in the works of the late biologist Stephen Jay Gould. He had a name for it: panadaptationism. Like Kipling's famous "Just-So Stories," some of our evolutionary stories are likely to turn out to be fables. Others are fabulous in a good way: richly explanatory as well as quite likely to be true.

A good example of a just-so story concerns the origin of the philtrum. This dip in the upper lip beneath the nose lets humans make more expressions and sounds than we would be able to without it; it thus probably has a connection to the evolution of communication and speech in our species. But how the human got his groove on has a simpler story in mythology. In the Jewish Talmud, God sends an angel to the womb to teach the unborn baby all there is to know. Prior to birth, the angel touches its lip to relieve it of its wisdom, leaving the indent of the philtrum. In this (and other) just-so stories, this human lip groove is the visible sign not of knowledge but of its divine removal, the "shushing" of a sacred secret.

Evolutionary stories are not so memorable or beautiful. While fascinating in their own right, the stories of how we

lost our hair to become naked apes are not as simple as the mythological tale of how we acquired the groove above our upper lip. The truth of our complex, reproductive, sex-enabled bodies and minds involves myriad organisms and multiple ancestors. We may have good ideas about how such and such an organism got such and such a feature but, unlike in myth, there is always room for doubt. We did not arrive full-blown from the pages of a book, but arose gradually over multiple generations along long and winding paths. When we study the evidence, we will still be left with questions. Self-understanding is not always flattering, nor is knowledge of ancestral sex lives necessarily conducive to pride. But the partial stories of evolutionary science, though not complete, are more than beautiful fairy tales. They offer reality as well as enchantment.

And that's the naked truth.

⁓❦) 3 (❦⁓

Monkey Traits

*S*uper-soft fur and slippery skin. Or is that lickable nipples and arguable kin? Or fun-filled frolicking in the name of sin? Whatever we call it, however high it flies on the rarefied notes of an aesthetic sensibility or low it sinks in the aftermath of familial responsibilities and limited options, the urge to merge—the lustful morass of feelings, emotions, and relationships around which mammalian sexuality swirls—begins and ends with bodies. To understand it, we must do a little time traveling. Fortunately, time travel itself is, so far, impossible. Fortunately because, if you were to go back and fall in lust with a fur-clad cave hunk or hottie, you might sire or give birth to a boy who grows to a man who kills your own ancestors. That would not only be a science-fiction paradox but also deprive you of the pleasure of reading this book.

But if we can't go turn the clock back, or depend on evolutionists' just-so stories, how *can* we find out what our ancestors were up to?

A powerful tool in reconstructing probable ancestral sex lives—less "just-so" than "might-be-so" stories—is comparative anatomy. By looking at now-living related organisms, we can see what traits they share and backtrack to determine probable features of an ancestor. The same can be done by comparing behavior, mating systems, and DNA sequences. There will be false leads, but, like the weight of circumstantial evidence carefully employed to re-create a crime scene, we can come up with a plausible picture. And unlike the prosecut-

ing attorney, a scientist does not have to prove his case beyond a reasonable doubt. The continuum stretches not between crime and punishment, but between curiosity and discovery. New evidence will not get anyone out of prison, but it may release us from the subtler incarceration of received opinion.

In the 1980s, and although in the center of a full house near the front row, I walked out of a lecture by a creationist who was trying to make fun of evolutionists during the course of his slide show. "Evolutionists want you to believe," he said, flashing a crude cartoon of a cow by the seashore, "that *this*"—and then our intrepid advocate flashed forward to a picture of a great whale in the water—"turned into *this*."

Ovid, in his *Metamorphoses,* recounts some startling transformations. But a cow turning into a whale is not one of them, any more than it is for evolutionists. Caricatures and straw men do not an argument make. It is true that the ancestors of whales, dolphins, walruses, and seals were likely land mammals—more like goats than cows but in truth neither. Embryonic humans resemble embryonic mice and chickens—all three in utero look literally fishy: We have gill slits and tails before we come out of our mothers. Why would a creator give us gill slits in the womb, unless he used evolution to create, or was a prankster?

Anatomical similarities often reveal shared evolutionary roots. The evidence of common lineage is not limited to embryos. It is literally in our bones. The foreleg of a horse, the wing of a bat, the flipper of a whale, and the arm of a Moulin Rouge dancer all share a similar skeletal infrastructure.

Even the more honest creationist tactic of finding God in the gaps in the fossil record—emphasizing missing links—misses the point: What is remarkable is not what separates, but what connects us. Like a giant jigsaw all scientists are working on concurrently, missing pieces continue to be found. And they are profound. The 1850 Berlin discovery of the winged

reptile *Archaeopteryx* would have delighted Darwin, in whose time the jigsaw puzzle, mostly due to the dearth of paleontological piece finders, had just begin. Today a slew of new fossils of feathered dinosaurs have been unearthed in China. Indeed, paleontologists now classify birds *as* dinosaurs: They lay eggs, have scales on their feet, and are technically reptiles. Paleontologist Jack Horner (an inspiration for the book/film *Jurassic Park*) even claims to be able to produce a modern-day mini dinosaur by interfering with embryonic development of a chicken, a small featherless dinosaur with teeth.

We are backboned animals with anatomical and sexual characteristics similar to other organisms that share our ancestry. The coccyx, the little tailbone at the bottom of our spine, serves no purpose for us now but it did when our simian ancestors swung from the trees. A grasping tail is an excellent tool if you are used to clinging to a branch as you call out for a furry friend. The great and lesser apes and Old World (African and Asian) monkeys all lack grasping tails. Some of the smaller New World (North and South American) monkeys, the smallest of which is the pygmy marmoset, a paltry lightweight at five ounces, have grasping tails. Unlike bigger Old World monkeys, the New World simians rarely come down to the ground, except for the occasional nut or cricket, preferring to scamper about from branch to branch (some, such as the marmosets, feeding directly on tree sap with special bark-piercing teeth) in the tropical forests in southern Mexico, Central and South America. Although it's impossible for landlubbers to keep full account of the sixty-odd species of New World monkeys, their sexual and social relationships vary, with, for example, male tamarins and marmosets (whose females typically give birth to twins) carrying the infants most of the time, whereas daddy capuchins (the famous organ grinder monkeys) do not tend to take care of their offspring; some New World monkey species have harems with one male and several female

consorts, while others, such as the callicebus monkeys (titis), tend to form long-term monogamous relationships. A similar variety marks the apes and Old World monkeys, who are more closely related to us.

The Platyrrhini, the ancestral stock that became the New World monkeys, may have arrived in South America on floating chunks of vegetation. They could have traveled on a natural raft like the floating mangrove forest islands that violent storms sometimes break off the coast of Africa. Geographic isolation—the separation of populations as the result of such events—was probably a major factor in the evolution of primates. A floating island, earthquake-separated patch of jungle, or primate tribe following fruit trees into a remote and distant valley and remaining there may separate members of a genetic stock. Physically separated, they no longer interbreed. Ultimately troops and tribes went their own way, evolving to the point that they could not form fertile offspring with members of the ancestral lineage even if they were still able and willing to mate with them. In this way new species, including our ancestors—who were mating long before there were humans—formed.

Genetic and fossil evidence suggests that the flat-nosed, branch-swinging New World monkeys split from the Old World monkeys—baboons, macaques, and many more—some forty million years ago. The island-hopping ancestors to the New World monkeys would have been aided in their journey on floating clumps of vegetation to the New World because Africa and South America were closer together thirty-seven million years ago in the Oligocene epoch.

The Old World monkeys, like us and apes, are catarrhines (Greek for "hook-nosed") with downward-pointing nostrils. The biggest superficial difference among the three great primate groups closest to us—the Old World monkeys, the New World monkeys, and the African and Asian apes—is in the

tails. The catarrhines, when they have tails, can't hang, clutch, or hug with them as can the broad-nosed platyrrhines. Old World monkeys and the apes, like us, despite some vestiges here and there, have outgrown them. This could be because, unused, any changes that shortened tails had no material effect on survival, as our Old World ancestors gave up navigating the arboreal jungle gym for splendoring in the grass. Use it or lose it. But the true tale of the tail, as usual, is probably more complex. The coccyx, uterine tail, and occasional birth of children with tails indubitably suggest that our ancestors had tails and that, if we are made in God's image and the devil an angel, they may also have been so endowed.

By looking more closely at the members of our evolutionary group, we can glean something of our shared ancestors' sex lives—the erotic ape matrix of which human sex lives, despite their variety, are only a perhaps passing variation.

The evolutionary family Hominidae to which humans belong includes two species of chimp, the common and bonobo; three subspecies of gorilla, western and eastern lowland and mountain gorillas; and two species of orang, the Bornean and Sumatran. Immunological studies in the 1960s showed that the African apes are far more closely related to us than to Old World monkeys.

Although not directly answering the famous barb of Bishop Samuel Wilberforce in his 1860 debate with evolutionist Thomas Henry Huxley as to whether it was through his grandmother or his grandfather that he claimed descent from a monkey, a combination of fossil, immunological, and genetic evidence suggests the Old World monkeys split from the great ape lineage of which we are part some thirty million years ago. Various methodologies suggest that the orangutan line split off from the other great apes about fifteen million years ago, the gorillas about seven million years ago, and humans from common ancestors with chimps some five million years ago.

A recent genetic study that offers a clue about the fur gap between people and the other hominoids has to do with a protein. As shampoo ads sometimes mention, proteins are a major constituent of hair. In fact our bodies are mostly protein—blood, skin, organs, toenails, hair, and so on are all made of proteins. The main sort of proteins in hair are called keratins. The journal *Human Genetics* suggests that one of these proteins, human type I hair keratin, appears to be coded for by a gene that may have been inactivated some time after the divergence of *Pan* (chimps) and *Homo* (modern and extinct humans). This gene is one of the eighty that have been lost—thirty-six of which code for olfactory receptors allowing a better sense of smell. A disproportionate number of the other genes lost had to do with immune response, perhaps reflecting different pathogens in the primeval environments in which we and our soul-sister lineage evolved. But losing the type I hair keratin gene may have been the immediate cause of human body hair loss. The massive thinning and loss of our ancestors' body hair is estimated to have occurred about 250,000 years ago, very recently in geological terms.

But behind the immediate genetic cause may well lie a deeper cause. Evolutionary biologists distinguish between ultimate and proximate cause. *Proximate cause* refers to immediate chemical or physical cause. *Ultimate cause* refers to evolutionary factors that can no longer be directly observed. One of the first to postulate an ultimate cause for human hair loss was the author Desmond Morris, who intriguingly suggested that sex was part of the story of why our ancestors lost their fur.

4

Affair of the Hair

A nymphomaniac likes to fuck; a trichophiliac likes to pluck." While not exactly a shining example of the political politesse requisite to pick up a debutante at a White House ball, I recently overheard the rhyme at close range. Forgive the profanity but, since it may help you recall the name for a practitioner of the second predilection—which is not confined to plucking but refers to hair fetishism of all kinds—I thought I'd pass it on.

The loss of hair in our lineage perhaps 250,000 years ago had fateful consequences for our forerunners. Indeed, it may be that, without the loss of body hair among the ancient Africans that were to evolve into humans, our kind would have developed neither close-knit couples nor babies with big brains. Having studied with the Nobel Prize–winning ethologist Niko Tinbergen, zoologist (and surrealist painter) Desmond Morris speculated in his 1967 best seller *The Naked Ape* that hair loss, which exposed the very sensitive human skin, would have increased our species' tendency for eroticism. Skin is the largest organ of the body; hairlessness would have expanded its zone. The enlarged area sensitive to touch, in turn, would have allowed for sexual sensations to expand. We could have moved from the rear-entry intercourse more common in the animal world to more communicative lovemaking sessions in which increasingly faithful primate lovers gazed into each other's eyes. This primeval penchant for face-to-face union, Morris reasoned, helped our ever more expressive, erotically

empathic ancestors to stay together for the special challenges of caring together for extremely vulnerable human infants.

Is Morris's a just-so story? Well, sometimes Morris's animated imagination did carry him into Aesopian fable territory. For example, he suggested that a woman's bare breasts in the savanna, by mimicking a pair of colossal eyes, might have scared away potentially dangerous predators. Lush female lips, he further speculated, mirrored the gently parted vulva, and would have enticed males to come around the front and look their lovers in the eye—as would rounded breasts, repeating the ancient sexual signal of paired hemispheres, on the other side. All this was part of a supposed transition that led, if not to true love, at least to more devoted partners better able to be attentive, fully human parents.

Although the story contains elements that may well be true, observations of the bonobo call aspects of it into question. Discovered in 1928, this endangered species of ape, *Pan paniscus,* is found only south of the Congo River in the Democratic Republic of Congo. As with the platyrrhines, whose ancestors may have settled South America after coming across the Atlantic Ocean on a storm-tossed island raft, so the origin of the bonobos may owe to geographic isolation. In their case, being poor swimmers, the appearance of the Congo River two million years ago cut the population from which they derived in two. The primates in the south evolved into the bonobos, while those in the north became the more populous common chimpanzee.

Compared with chimpanzees, bonobos are known for their smaller size, longer legs, hair parted in the middle, facial individuality—and their sex parties in the trees. Rather than dragging their knuckles, like common chimps, they walk upright about a quarter of the time, and when they do stoop-walk they use their palms. Their lips are pinker, their faces smaller than chimps; they have less prominent brows and longer head

hair. Bonobo babies laugh when tickled. It has been suggested that there is something more than a little human about them. Emory University primatologist Frans de Waal of the Yerkes National Primate Research Center in Atlanta tells of the amazingly human-like behaviors of chimpanzees. One would refuse bananas and only come down from the roof if his caretaker began weeping, in which case he would descend to put his arm around her in consolation. Some primates gladly perform tasks for cucumber slices until others are given better-tasting grapes, at which point they throw down their cucumbers in protest.

Indeed, on the basis of a 2003 study showing that 99.4 percent of certain DNA sites are identical in corresponding human and chimp genes, Morris Goodman of Wayne State University proposed that both common chimps and bonobos be classified as members of our genus, perhaps *Homo paniscus*. (The title of this section, Humans and Other Chimps, is not meant literally, but as a provocation and palliative against our strong tendency to consider ourselves separate from, and superior to, other primates.)

Bonobos are female-centric. Other primates, such as squirrel monkeys, whose males greet one another with erections, use sex for communication, yet none are such politic libertines as the bonobos. Sexual intercourse and oral sex are used in greeting, conflict resolution, and reconciliation—which would be the bonobo equivalent of "makeup sex" if they hadn't been so liberal with their love as to spread it around their literally high society: In couples, threesomes, and moresomes, adults engage in high-flying tree sex, babies swinging from nearby limbs or even the arms of their randy parents.

Generally kinder and gentler than the common chimpanzee, bonobos eat less meat. Although they do occasionally hunt other primates, such as redtail monkeys and Wolf's monkey (*Cercopithecus wolfi*), when they make a kill they share

it. Celebration of a new food may lead to an orgy, and food, like sex, is shared rather than hoarded. Females have more social status than in common chimpanzees and other apes, the bonobo's status deriving largely from his or her mother rather than powerful males. Females are smaller than males individually but their erotic partnerships, revolving around the practice of mutual vulva rubbing, make them socially dominant.

The reason bonobos pose problems for Morris's ideas is that, despite their hair and lack of male–female parental pair bonding, they can be quite good, attentive lovers, adept at face-to-face communication. They engage in extended bouts of tongue kissing and "missionary position" sexual intercourse. They have managed to "move round to the front" without giant breasts, and they have an extremely promiscuous society without the expansion of the erogenous zone due to loss of body hair.

The loss of body hair in our ancestors was probably not directly related to touching or exclusive relationships. An alternative scenario is that hair loss, when it occurred via mutation, was strongly selected for because it led to more effective heat loss. In a hot climate the ability to lose heat can be a big advantage, allowing you to move faster for longer periods of time. Even standing up, versus walking on all fours, would have tended to cool primates—both by exposing less of their surface area to the sun, and by accessing cooler temperatures and breezes. Body hair loss would have helped even more. Organisms that lost hair—perhaps through the loss of the gene for keratin I hair protein—would have enjoyed a great advantage in regulating their temperature—cooling off—in tropical and subtropical temperatures.

These cooler ancestors would have had a big advantage. Most people don't realize that humans are the fastest long-distance runners in the world. Over shorter spurts, the 70-mph

cheetah is the fastest land animal. But over long distances no one can outrun marathon-running human beings.

Lack of hair allowed the skin to sweat without obstruction. This in turn led to enhanced evaporation—a major breakthrough in body temperature regulation in a tropical climate. The runners, running farther, could catch up with their prey. They accessed more fruit trees over shorter periods of time. But that's not all. Thirty percent of the blood sugar that powers our animal metabolism is used by our natural information processor, the brain. Food-accessing, decreased-fur primates, covering more ground, would have procured more fruit sugar and other nutriments running not only their legs but also their thinking. The greater food accessed by running created a cycle involving greater blood flow, more energy procured from the environment, and more energy available to plan more hunting, fighting, and loving. So cooling via hair loss and evaporation could plausibly have initiated a cascade of benefits to our increasingly naked ancestors.

Another possible piece for the hair loss puzzle was advanced by feminist Elaine Morgan in her book *The Descent of Woman*. Morgan made a splash, so to speak, with her presentation of the aquatic ape hypothesis. Critical of Morris's macho-oriented depiction of newly hairless males chasing game across the grasslands, she wondered what the advantages of reduced fur would be to females, who were depicted as staying behind gathering nuts and roots. She contacted Morris about something he had mentioned, Alister Hardy's thesis that humans had undergone a time of evolution in the water. Many hints pointed in this direction. We have a layer of subcutaneous fat, characteristic of mammals such as walruses and whales that have returned to the water. Foraging in the water would also have entailed keeping one's head out of it, which would have helped in standing upright. Bears and other mammals that feed in the water are more likely to stand upright. Even

the philtrum, the groove above the upper lip, allows some people to block their nostrils, preventing water from getting in their nose when swimming. Then again we need omega fatty acids in our diet—available, aside from relatively hard-to-get or obscure sources such as flax and linseed oil, only from fish with whom we may have shared some time in the ocean.

Hardy, whose original epiphany centered on the reason for blubber in humans, sat on his ideas for thirty years so they did not interfere with his entrance into the Royal Society. But Morgan, a Welsh housewife and writer for the BBC, consulted him and extended his ideas, suggesting that babies in the water may have grasped their mother's long hair. People today are often born with webbed toes. The human body appears to be hydrodynamically streamlined, with the ability to swim underwater like a frog. In relatively modern times such swimming has been used by sponge and pearl divers holding their breath for a long time under the water, but in the past it could have been used to enable capture of burrowing crabs and shellfish, bivalves and sea urchins. Fishing with our bare hands would have helped predispose our ancestors to manual dexterity, while breath holding could have helped kick-start the origin of spoken language as well as, unfortunately, sown the seeds for long-winded political speeches. Hardy points out that other sea mammals such as whales, manatees, and hippopotami have lost their hair, that the main place where hair remains on humans is at the top of the head where it would stick out of the water, that hair underwater loses its function of keeping the body warm, and that, unlike other primates, the minor hair that remains on our backs points diagonally toward the body's midline, exactly where streams of water would pass around the body to meet in its forward swimming motions. Such hydrodynamic arrangements of hair, and loss of body hair in general, would have lessened resistance to swimming bodies. For comparison, the dolphins and whales

possess vestigial remains of hind legs tucked into their boat-like bodies. Hardy and Morgan envision a time when competitive life in the jungle drove our glistening forebears to the shore, where they waded into the shallow seas grabbling with their bare, fin-like hands for sushi.

The theories of Hardy, Morris, and Morgan aren't mutually exclusive: Losing our hair in the water would have let us come away prepared to be much better runners. Hair, which works as a heat conductor by keeping air stationary between the follicles, would have been useless insulation underwater. Then again, all that experience wading would, like a toddler being supported by its parents, set us in an upright stance, priming us for hunting, dancing, and miniature golf. The expansion of the erogenous zone would have been a side effect. Our lack of hair, combined with the loss of genes connected to smell, makes us more touch-oriented than other mammals. You don't say you haven't smelled someone in a long time, you say you haven't seen them. And ask them to keep in touch.

Which brings us back to this essay's opening gibe. Many primates spend a lot of time grooming one another. They sit on a branch, or on the ground, and methodically pluck the insect parasites out of one another's hair. This has a survival benefit. Some of these insects carry diseases, some of which are fatal. Having a partner or friend pluck out your parasites is also pleasurable, the chimpanzee equivalent to a back tickle or massage. Although more associated with women, men are no strangers to hair care. (Even that archetypal macho loner, Clint Eastwood, seeks respite in the caresses of the barber chair.) Since we have lost our hair, it is possible that, beauticians aside, we instinctively miss grooming one another. Losing hair got rid of lice and ticks, stripping them of some of their prime primate real estate. But it also left us, evolutionarily speaking, out in the cold. I wonder if the young woman who, feeling isolated in the modern urban setting, picks at her skin is not

instinctively trying to comfort herself, as if retreating to an earlier evolutionary time when our ancestors had more hair, and the plucking of parasites was among a primate's most pleasurable and profitable pastimes.

Being groomed, having one's hair cut, like a massage, caresses, or laughter, can produce endogenous endorphins, the body's pleasure drugs, its natural "opiates." Indeed, although tearing one's hair out is itself an idiomatic indication of being totally stressed, hair tugging, tousling, and scalp stimulation in general release endorphins. Technically *trichophilia* refers to amorous longings for that part of the body that is hair. It is not the most common fetish, coming after, in a recent Italian survey conducted at the University of Bologna, longings for boots and underwear but before having a thing for muscles or tattoos. Still, evolutionarily, it is one of the kinkiest.

$\text{-}\!\odot)$ 5 $(\odot\!\text{-}$

Secret Races

$\mathcal{W}$hy is a promiscuous man a "stud" but a loose woman a "slut"?

Seekers after the source of the varied value judgment put on multipartner men versus multipartner women need to look at the crossed currents of the mating lives of our esteemed ancestors. Our mating system is distinctly "mixed": delicately balancing the bullying possessiveness of gorillas with the swinging promiscuity of the chimp. Although, unlike these close relatives, we tend to form loving couples, monogamy ("monotony") is likely of relatively recent evolutionary vintage. It came into being because females who failed to tend to vulnerable babies whose brains developed outside the womb lost them. If they had devoted help from a male, who might easily be enticed, at least at first, by the lure of an available sex partner, the chances of rearing healthy offspring to fruitful maturity were much increased.

Although recent studies shed doubt on the once much-bandied-about figure that we share more than 98 percent of our DNA with chimps, there is still broad agreement that of all our planetmates, these are the most like us. And bonobos—with their longer legs, pinker lips, more distinctive faces, longer head hair parted in the middle, laughing babies, and greater tendency to walk upright—still more so. Compared with chimps, the other great apes are prudes. Orange-haired orangutans (from the Malay and Indonesian, meaning "forest people") are palm-walking loners inhabiting the islands of

Borneo and Sumatra. When they meet a mate, they couple but, afterward, return alone to their trees. The largest tree-living primates, they fiercely defend their territory. Immature males will mount any female, although strong older females fend them off.

Gorillas have harems, ruled over by dominant males called silverbacks, distinguishable by the patch of white hair on their curved lower back (which correlates with their age and status). A silverback uses his strength and status to mate any females in his troop. They are "his." Adolescent males resort to ruses such as seducing females out of the line of silverback sight, and they may team up to try to oust him.

Common chimpanzees go into heat, or estrus (from Latin *oestrus*, "frenzy")—a period of sexual fertility and receptivity that is clearly demarcated by a (to-us) ridiculous and almost scary-looking swollen pudendum, bright pink in color. Male chimps love this look, though. An estrous female will attract virtually any male and mate with almost any male. When industrious chimps lead fertile females into the woods "on safari," they are physically reprimanded upon return from their illicit "honeymoons": The remaining males, who tend to be genetically related, box the secret lovers about the ears as punishment for their looting of precious booty.

The rough treatment of the lovers reminds me of a story. I heard it when I was twelve at Camp Kinderland, a New York–based communist-inspired camp (Paul Stanley of Kiss, and Bobby Fischer, the chess player, had attended), then located in the Poconos Mountains of rural Pennsylvania. Years before I arrived, a teenage couple were caught fornicating in the showers. As their punishment, the miscreants were forced to march naked throughout the camp. It was hard for me to imagine this quirky place of puppy love and dancing to Motown on summer nights being so punitive. Maybe it was their communist mandate. After all, they were ideologically Marxist enough

to force us campers to share any candy we got from our parents in the mail or from surreptitious trips to the nearby Pennsylvania town. And, though Plato was not much of a communist (Diogenes said, "Offer Plato a fig, and he'll take the whole bowl"), *The Republic* expresses similar ideas about sharing women. Thinking back now, I wonder whether if behind the punishment meted out for the "crime" of premarital sex there was not a deeper implication that nothing, not even young love, should be hoarded. The chimpanzee couple that head off for a sexual tryst are also scolded by their society, far more communist in its way than the guardians of Camp Kinderland, who were after all probably just enforcing, albeit with callous creativity, some variant of the camp's conduct code.

Female chimpanzees, upon reaching adolescence, leave their natal troop. They "run away" from home to join another troop of related males. When they become fertile, pink and genitally swollen, physiological signs of ovulation, the social expectation is that they are fair game for any male.

The mating differences and similarities among gorillas, chimps, and us reflect secret races and ancient contests that began long before the human species arose. These X-rated contests reward the animals, by way of continuance of their genes, who are better able to impregnate females or prevent others from doing so, who produce more sperm, or ejaculate farther, or more often. Rewarded even are males with especially sticky sperm that coagulates on contact, as it does in some rodents who make "sperm plugs" that, like natural chastity belts, form a barrier blocking subsequent suitors. The same is true of the honeybee mentioned earlier, whose detached penis blocks new suitors. Some species of dragonflies and most damselflies have ornate penises (species are told apart by them) with various sorts of scoops, scrapers, flanges, hooks, horns, and curved spines that they use to remove previous sperm from a female's spermatotheca and bursa copula-

trix, which are special sperm storage organs. The typical male damselfly spends most of his time during copulation removing previous males' sperm, introducing his own only after cleanup makes it evolutionarily worth it.

The sperm contest rewards seductive males and good lovers who provide females the momentary pleasure of orgasms—which sexological studies show produces an intrauterine vacuum, thereby drawing in sperm and increasing the chances of some winning the ancient competition. And of course, the males who win these fertilization contests are more likely to sire offspring whose genes code for the same sperm-contest-winning traits that fostered their own illustrious existence.

But the secret race only really starts when females are promiscuous. If females are completely faithful to only one male, there will be no race, as there will be only one contestant.

This effective way to win the sperm race—to not let other males play—is the gorilla's harem strategy. Using brute force prevents competitors, no matter how pornographically large their penises or voluminously large their testes, from reaching the starting gate of the vulva. Effective delivery of sperm to win the race can be negated by powerful alpha males. The human mating system is "mixed" because we show genital evidence of being formerly more promiscuous, but we also have sexually possessive males, patriarchy, and male domination.

When genitalia of the great apes are compared, the chimps win. The average chimp produces 650 million sperm cells per ejaculation, the average human 200 million (although sperm counts have been going down, probably because of pollution), and the average gorilla 65 million. Humans do have, on average, the biggest penises. Gorillas, despite their imposing physique, have erect penises that measure on average four centimeters. That's an inch and a half long, surprisingly small for King Kong. In some African languages, "hung like a gorilla" is an insult. In terms of penis-to-body ratio barnacles, whose

penises are many times the length of their sessile bodies, win. The average sea lion erection is five inches, human six inches, stallion, rhinoceros, and walrus two feet. A bull's is three feet, an elephant's six feet, and a humpback whale's weighs in, or rather stretches out, at ten feet.

In the males of promiscuous primate species the greatest advantage, in terms of siring offspring, tends to go to those who produce the most sperm. The reason is not so hard to understand. Because estrous females are willing, and able, to mate multiple times with virtually all available males, except their own sons, those males with more sperm have higher chances of being fathers. Male chimps are only about a quarter the weight of male gorillas, but their testes are four times as heavy. This makes sense when you compare the different mating systems of these two species. An estrous chimp may mate with sixty males in a day who, whether they realize it or not, are thereby competing to impregnate her. The proverbial chest-beating gorilla, by contrast, attempts to use his superior size to intimidate other males (usually adolescent males) from erotically approaching any of "his" females.

Scientists refer to all this as sperm competition. When Darwin emphasized the shaping characteristics of sexual selection, he was thinking about males fighting for female favor at the level of their whole bodies—antlered stags, enraged elephants, and dueling aristocrats—rather than the subtler contests of promiscuous sex. Males sometimes fight to the death for female favor. This, in a way, is sperm competition. More correctly, it is sperm competition *avoidance*. It is called avoidance because, like a giant bouncer barring entrance to a ladies' man at a sexy nightclub, competitive male genital features such as more copious sperm production become irrelevant if the male cannot enter in the competition in the first place.

Darwin missed, as far as we can tell, this sexual selection that occurs when females mate with more than a single male

during a fertile period. Only after Geoffrey A. Parker, a zoologist and amateur jazz clarinetist at the University of Liverpool, elucidated the concept of sperm competition in papers during the 1970s did the importance of sexual selection among sperm cells and their associated delivery apparatus become apparent.

Perhaps Darwin thought of it, but neglected to mention it because of Victorian scruples. But sperm competition shapes sexual bodies. Indeed, if we travel far enough back in time, it seems clear that anything in the direction of a penis would have been selected for: It allowed males evolving such instruments to preempt their competitors. Fish and most marine animals spawn by releasing sperm over eggs in the water outside their bodies. This is a messy system that is "improved" if a male can get to a female's body before she releases her eggs. This seems to have been what happened in evolution. The evolution of penises and internal fertilization, which occurred in our lineage long before we were apes, was most likely spurred by sperm competition. Despite the work done on this subject beginning in the late 1960s, it took decades for it to surface in the zeitgeist. In fact, in 1990 in New Orleans, through interest and connections, not expertise, I organized and introduced the first scientific symposium of the American Association for the Advancement of Science devoted to the subject: "Pollen and Sperm Competition: The Importance of the Haploid" (*haploid* referring to sperm and egg having only half a full set of chromosomes). Pollen tubes, which grow from pollen grains and transport the immobile sperm cells in many plants, also compete, racing, albeit in relative slow motion, to get to the female reproductive tissue.

To sum up, sexual competition occurs not only at the level of whole bodies before sex but, in promiscuous species, among genitals and sex cells after copulation. Position during intercourse, force and timing of pelvic thrusting, number and swimming speed of ejaculated sperm, and proximity of the

spermatic means of delivery to the egg at time of ejaculation all may favor one male over another in this ancient race that for millions of years has shaped our bodies and influenced— some might say warped—our minds.

For sperm competition's effects are not just physical. Swiss researchers have demonstrated that male stickleback fish that are shown computer-animated films of male rivals ejaculate more sperm when a spawning female is put in their tank than if they had not seen the aquatic porn. But it is not just sticklebacks. Studies show that men who suspect their mates of cheating on them produce measurably more sperm. The male body appears to have an unconscious compensatory mechanism, upping sperm production if it detects a potential for rivalry or cuckoldry. This leads to a curious relationship between jealousy and sexual excitement. In the famous novel *Venus in Furs,* Leopold von Sacher-Masoch's hero, Severin, is told by his beloved widow Wanda that he will not be her only. "I believe," she says, "that to hold a man permanently it is vitally important not to be faithful to him. What honest woman has ever been as devotedly loved as a hetaira?"

"There is a painful stimulus in the unfaithfulness of a beloved woman," replies Severin. "It is the highest kind of ectsasy." The aroused but alarmed Severin signs a contract to be the sex slave of the redhead with jewel-like green eyes. Soon he is gathering information for her about the men she wants to date, listening to her lectures on his vows of slavery and her freedom, submitting while she kisses away the tears in his eyes before whipping him again to recapture for a painter her look of beautiful cruelty, and so on.

After reading the book Sacher-Masoch's future wife, Aurora Rümelin, in her twenties, living nearby in Graz, Austria, with the help of an older female, wrote a shameless letter under the name of Masoch's heroine, Wanda von Dunajew. At first, she hesitated, then sent the letter. He answered his fan imme-

diately; he had read her letter in a rapture, he said. They met, and were married.

In her memoirs Aurora von Sacher-Masoch writes of being a harried housewife who, after household chores and taking care of the children all day, had to put on furs. Her husband also set up many opportunities for her to cuckold him, but she wouldn't do it. Not only would she not do it, but her refusal to hurt him hurt him—her failure to cheat on him was a major cause of their divorce.

Their relationship was thus a real-life version of the old joke:

Q: What did the masochist say to the sadist?
A: "Hit me!"
Q: What did the sadist respond to the masochist?
A: "No."

So, too, the novel *Story of O* (which started in real life as a series of letters written by Anne Desclos to her married lover, Jean Paulhan) features the willing submission of O (Odile, but also, by extension, orgasm, orifice, and the woman as pure sex object) to the sexual desires of many men. O agrees to the treatment, much as Severin signs a contract saying he will have no recourse but to do what Wanda says, up to and including killing him.

Such scenarios, in which female promiscuity is expressly permitted, are no doubt psychologically disturbing, but physiologically they can lead to greater male passion and sperm volume. Under conditions of female promiscuity, the gates go up on races more ancient than man or monogamy. And if jealousy can incite anger and violence, it can also be an aphrodisiac.

Blue Light My Baby

*L*anguage is musical. If sperm competition accelerated the evolution of internal fertilization and genitalia, so sexual selection may be behind music and, through music, language. Unseen in the night or in broad daylight crickets, bullfrogs, and birds are among the many animal lovers that advertise their presence and send sonic notes to potential mates or parental partners. Great apes also may have made mating calls. Before meaningful speech they could have made beautiful music together. From music, some evolutionists speculate, language itself could have come. As Darwin himself pointed out, language may have evolved via sexual selection as lovers calling out to one another competed in the aural realm.

Language is a window into the mind. I know what you think, or think I do, if I hear you. Complex sounds, at first playful and then meaningful, would have attracted primeval lovers no less than modern musicians do—not out of frivolity, but because musical and lyrical skills can be spellbinding displays (accentuated in musicians' cases by sometimes wild clothing choices) of primate intelligence. In a world where predators can eat you, members of your own species can kill you, and problem-solving skills can save you, intelligence is a desirable trait. Females who mated with primordial musicians may have given birth to cleverer, more likely to survive offspring.

Charles Darwin called the origin of the flower an "abominable mystery." But for his colleague, the naturalist Alfred Wallace, brains were a bigger mystery. Unable to see how big

brains could be explained by natural selection, Wallace went so far as to consider them evidence of a creator. But it may have been sexual selection that got us our big heads.

Casanova was a soldier and priest as well as a world-famous lover who disdained mistreatment and loved laughter and food. A friend of Mozart, he did not consider himself well endowed, and preferred affairs over one-night stands. Why mention him? Because Casanova, with his many talents—he played violin professionally in Venice's San Samuele theater—and his many illegitimate children is a paradigmatic example of an attractive man captivating females with his "musical mind." The superfluous verbosity, the primeval sex-tinged babble from which language may spring gives an indication of the inner characteristics of the sound maker. Like words on the Internet, sounds are a form of appearance by which mates can be judged (though not always accurately). If eyes are windows to the "soul," so sounds are portals into the mind and its thoughts, the thinking brain and the genes that helped form it. Our genetic hearts, in other words, cannot be fully estimated only by taking measure of our bodies, however lovely or odious, because intelligence and the survival abilities it enables, especially in the hominid line, are more than skin-deep.

Nonetheless, we may have inherited our big brains—and consequent qualities ranging from art to paranoia, entertaining to warmongering—not because they had immediate added survival value, but rather because women, or the humanoid ape females that preceded them, were on the genetic equivalent of a shopping trip, where their choices weren't clothes or perfumes, luggage or cosmetics, electronics or baby toys but, rather, men. And their choices tended toward the musical and verbose. Not only because better communicators were better lovers and more able to survive, but because . . . well, just because—because speech-like sounds and musical tones were in primeval fashion. Which would explain why we have

linguistic and artistic abilities above and beyond those needed for survival. And why women go for musicians.

Such a scenario is an example of a "runaway" process in which certain sexually attractive characters become exacerbated. Modeled by the famous mathematical evolutionist Ronald A. Fisher in the 1930s, the basic idea is that there is some trait that correlates with higher "fitness"—that is, the ability to produce more, and healthier offspring. A preferred trait, such as brilliant plumage in peacocks, becomes rapidly selected as it is sexually preferred, appearing in the progeny of those who select it in the next generation, thereby begetting a feedback loop that increases a desirable trait until it reaches exorbitant levels—even levels, such as the heavy antlers of stags, where it interferes with natural selection proper. This may be the case with our propensity for speech.

Chimpanzees are far less gifted than our species in the vocalization department. An extensive review of modern linguistic theories by Belgian scientist Mario Vaneechoutte and British neuroscientist John R. Skoyles compellingly argues that it was indeed the capacity for verbal imitation and music that led to language in human beings. Songbirds sing; parrots copy sometimes quite complex sounds—ringing phones, swear words, even their owner's address when lost. After bonding, animal couples in various species may maintain long-term contact by signaling musically, whether the sounds make sense or not. Vaneechoutte and Skoyles reference an article in *Nature* suggesting that facial bone structure differences between us and Neanderthals can be traced to evolutionary changes related to the acquisition of speech. But for Vaneechoutte and Skoyles it was not speech itself that was selected for, but rather musicality—the sexually selected ability to chatter playfully, repetitively, and imitatively—primordial entertainment before the invention of MTV.

Chimp faces look different enough for researcher Jane Goodall to tell them apart and give them names. And they can vocal-

ize distinctively. In one troop primatologists observed that, after one chimpanzee with his own typical call died, another took over the chimpanzee's typical sound. The special sound, absent from the troop since the chimp's death, enjoyed a revival—another chimp copied it. Skoyles, who himself was once diagnosed as communicatively impaired—and published his first paper in *Nature* when he was a teenager—points out that the chimp call mimesis is equivalent to the independent origin of a proper noun, a name. Monkey hear, monkey call: Repeating the call brought the physically departed chimp back to sound, if not life, and shows the sort of naturalistic routes that lead easily from blathering verbosity to meaningful language. Chimp chatter shows, as if through a glass darkly, or via a scratched phonograph record, how serendipitous sound-making activities can lead to reference, to invocations, to the magic of names.

This kind of naming could go along with more complex social thinking. It would not be long before other objects, which did not make sounds, could be given names. From the babble of babies, the squeaks and squawks of the scared, from the flirtatious overtures and arguments of adults, language in earnest could have evolved. As soon as it arose it could have given powerful new abilities in battle and hunting to the groups of sexually reproducing animals whose sound signals helped them recognize, impress, and keep track of one another.

A semiotic universe of technological significance and aesthetic merit may have been spawned by the blithering jabberwocky of the magical, musical, prehuman mating machine. Some scientists would like to drape a red celebrity rope around culturally evolving, speaking humans to separate them from the dumb animals. Only we, and not all of us at that, would be allowed into the real language club. But such separation violates the spirit of evolution. Sharing a common ancestry suggests that language and musicality are not exclusively human but rather, as Darwin implied, part of a continuum.

Pity not the slightly smaller chimp brain that cannot plow through a rhyming dictionary or devise bewitchingly lascivious lyrics with which to seduce the susceptible chimp chick. He knows not what he knows not, and it may, as in Eve before the fateful chomp, be a blessing.

If humans are more loquacious, we are not worlds apart from the chimpanzee. A great example, which I also take from Skoyles, illuminates the deep roots of our sexism. While it is undoubted that we are culturally conditioned, our civilized behavior has animal roots. Nowhere is this better illustrated than in the behavior of chimpanzees toward their mothers-in-law. Of course, chimps don't have mothers-in-law. The lickerous chimp female, taking on all comers, has not a clue who the father is. Nonetheless, chimpanzees recognize one another, and the mothers their offspring.

When a chimp grandmother comes to visit her daughter, whom she knows, and her daughter's offspring, for whom she naturally has feelings—she is not welcomed. The related males of the troop turn her away, harass her, and sometimes beat her. Because adolescent females leave their natal groups to seek out greener savannas, their visiting mothers come from another group. From an evolutionary point of view it might be of considerable benefit if the chimp moms could be helped, with offerings of food and care, by their own mothers. But the local males won't have it. They have no interest in strange females, unless they are fertile. With no chance of displaying the garish pink markings of estrus, or putting out the scintillating pheromones that drive chimp males into a frenzy, the mothers of the mothers of the males' offspring are not treated with even remedial respect. As is sometimes said about human males, the chimp males appear only to care about one thing.

Unlike us, the smaller-brained chimps apparently have no clue that they are fending off the grandmothers of their own offspring. Their free-for-all mating does not allow anyone to

know who the dad is. The males know not what they've done, or whom they've sired. Compared with us, they're socially retarded. Yet not knowing who the father is occurs among us, too, as does the sexual default position of males whose minds run obsessively to a single subject. The social monkey mirror they hold up is amusing. In it we see ourselves slightly changed. Men are no stranger to the ancient antipathy toward the mother of the mother. But in us it has been repressed. It reappears in muted form, in the socially acceptable realm of mother-in-law jokes.

If you doubt me, consider the following.

> Q: What is the ideal weight for a mother-in-law?
> A: About 2.3 lbs, including the urn.

> Two men were in a pub. One says to his mate, "My mother-in-law is an angel." His friend replies, "You're lucky. Mine is still alive."

> What should you do if you miss your mother-in-law? Reload and try again.

Most of the jokes are like this, centering on the absence by death of the mother of the mother. But not all.

> A guy was out shopping when he saw six women beating up his mother-in-law. As he stood there and watched, her neighbor, who knew him, said, "Well, aren't you going to help?" He replied, "No. Six of them is enough."

We've come a long way, but not as long as you might think. Although women today do not come into estrus, it is likely that the females ancestral to both humans and chimpanzees

did. And today we naked humans can mimic ancient primate body signs by wearing clothes. Not only did our ancestors chatter and drum and haltingly say one anothers' names, they also donned animal skins, adorned themselves with plant pigment, and made cave paintings. The human ability to change appearances, to create sensations, to signify is remarkable, so much so that some would say it—like clothes wearing, along with speaking—catapults us out of the jungle of animality into the God-given realm of civilization. But what is a "hot" girl, scantily clad and dressed provocatively in pink, if not a modern ape throwing off ancient signs of ovulation?

Chimp males have been recorded offering females meat in return for sex. Frans de Waal tells of wild chimps who risk life and limb to raid papaya plantations, coming back when successful and offering their bounty to fertile females in exchange for you know what. They call prostitution the oldest profession for a reason. Women don't have estrus, the vulval flushing and swelling that is an external sign of being in heat. Some sperm competition theorists speculate that humans went through a period of "concealed estrus" that worked to make males more faithful because they could never tell for sure who was ovulating. But experiments show a peak in sexual desire and attractiveness at the time of ovulation. Striptease artists receive more tips when they are ovulating, studies show. Sex workers, like all of us, can manipulate their appearance via clothes. If a beautiful escort walks past, strutting her stuff on high heels, heads will turn. If a streetwalker walks by in bright pink hot pants, she will attract even more attention. The color is no accident, I wager, any more than is the shape of the textured heart inside the fold of a perfumed Valentine's Day card. Such symbols resurrect estrus, long gone from the body, but ever open for business in the red-light district of the mind.

The Beautiful, the Dangerous, and the Confused

As often as not, it seems to be assumed that man has his being independently of his passions. I affirm, on the other hand, that we must never imagine existence except in terms of these passions . . . We are discontinuous beings, individuals who perish in isolation in the midst of an incomprehensible adventure, but we yearn for our lost continuity. We find the state of affairs that binds us to our random and ephemeral individuality hard to bear. Along with our tormenting desire that this evanescent thing should last, there stands our obsession with a primal continuity linking us with everything that is . . . this nostalgia is responsible for . . . eroticism in man.

—GEORGES BATAILLE

The Last Pornographer

*I*f we, as humans full of tall tales and musical dreams, can come up with all manner of story and associations, how are we ever to get to the bottom of the mystery of sex? Evolution, one avenue of solution, is certainly the main road taken in this book. But that, too, can be twisted. The mating mind is so full of sexual thoughts and erotic innuendo that it can see sex everywhere. Forget all the jokes that have double entendres and think of the more basic fact that the romance languages ascribe a gender to virtually every noun. In Spanish, the earth, *la tierra,* is feminine; the sky, *el cielo,* masculine. Never mind that the sun is male in some languages and female in others, or that in Ibiza and the Azores—and all Spain and Mexico and South America—*coño,* the vagina, takes the masculine article, *el coño,* like *el toreador,* the (male) bullfighter.

Sometimes the clearest example of a phenomenon is rendered not by the mundane version, but by the extreme case. Philosopher-novelist Samuel Butler said of Victorian society that there was nothing the English would like more than to see the fruitful union of two steam engines. American astronomer Thomas Chamberlin spoke of the sex-like merging of planets in collision. And comedian Lenny Bruce noted that men are not so picky; they will do it with mud or Venetian blinds.

Gendered thinking reflects our bodies. Even when we don't think we do, we have sex on the mind, as Freud showed. But

when it comes to coming to grips with the mundane omnipresence of thoughts sexual, few subjects are as strangely illuminating as the extreme case, that of French theorist Georges Bataille (1897–1962). Deeply influenced by the Marquis de Sade and Friedrich Nietzsche, Bataille was author of the infamous novella *Histoire de l'oeil*, The Story of the Eye. The book details the depraved adventures of a teenage farm boy and his girlfriend, Simone, who has a fetish for eggs and egg-like objects that leads her at one absurd point to insert bull's testicles in her vulva and, still unsatisfied, replace them with the bleeding eye of a freshly killed matador. The founder of surrealism, Andre Breton, expelled Bataille, who also influenced the literary theorist Blanchot and the philosopher Derrida, from the ranks of the surrealists, in whose social circles he traveled in the 1920s.

Like himself in his theoretical excess, Bataille's characters are exhibitionists. The young couple fornicate and urinate in front of the girl's aged mother; they seduce a beautiful, deranged neighbor in an impressively narrated lightning-storm scene. The deranged neighbor is committed to a mental institution, from which they help her escape only for her to kill herself. With a compulsive, if demented, logic, they have sex next to her dead body. Then they flee to Spain, where they seduce a priest in a church, forcing him to use bodily rather than symbolic fluids in a mockery of the eucharist; it is also in Spain that they witness the goring of the matador, whose organs the demented Simone refuses to let go to waste.

Roland Barthes argued that *L'histoire* is not mere pornography. It does not just string together a series of sexual episodes, but is a symbolic realization of the transgressive, other-connecting sexual mind. Bataille, in a postscript to his tale, admits that parts of the book were autobiographical, that he was conceived by a blind father and jerked off naked at night next to his mother's corpse.

Bataille's sexualization of reality is a study in perversion (from the Latin *per,* "away," and *vertere,* "to turn"—thus "turning away"); it is a perversion, for example, a turning away and application of sexual energy to the wrong objects, when, in one of Sade's fictions a son-in-law who promises to be chaste with a virgin bride-to-be and stops short of satisfying himself in the one way he most wishes—a Sadeian joke as, given the groom's desire for anal sex, the virgin will quite likely become pregnant even as he exercises his would-be erotic prudence. But it is also a perversion of a perversion: Bataille's insistence on transgressing, overturning, sullying, and sexualizing everything, while it is an affront to society and the proper, is also a perversion of the hypocrisy of preached but disobeyed moral strictures. It is a turning away from society's turning away from sex. Therefore it becomes an embrace of the sexual, of excess as the milieu of the sexual, as that which by nature does not keep its colors within the lines, does not stay neatly within its boundaries. For Bataille nothing was impregnable and everything was impugned. Sex, like death, was a pathway to transcendence, to the unveiling of artificial boundaries, including those of rational thought and the disciplinary boundaries of academia. He stepped through the doorway of excess to rub shoulders with the infinite. He used writing not to make deductive arguments but to make fun of them. He wanted to come, not to a conclusion, but to ecstasy.

In one of his essays he presents human evolution as a sexual phenomenon. Like an erection, primeval ape-man rises from a stooped to an upright posture, standing erect on Earth's surface. But the process, says Bataille, has not reached its logical conclusion. Our eyes still face forward, looking out parallel to the ground. Logic (which, as a proto-postmodernist, he was making fun of) dictates that for the process to culminate, our eyes should migrate to the top of our heads, merging to form an aperture through which we might ejaculate the contents of

our bodies in an obscene cascade toward the true object of our desires, the sun.

Like the gendered articles of unsexual things in Latin-derived languages, Bataille's extreme fantasy shows our tendency to project our own animal sexuality onto a more-than-sexual world. Yet as Barthes intimates, there is something more than sexual, something of literary merit, in Bataille's depraved imaginings. It is as if he were showing us a real quality of the universe that we had somehow missed. As if we could only take in the philosophical ideas he had to offer by presenting them in the form of sexual imagery.

In the recent movie *Sunshine,* directed by Danny Boyle, a crew in 2057 is dispatched to space on a dangerous mission toward the sun. They must adjust a nuclear imbalance in the sun that, if left unaddressed, will lead to its death. As in Bataille's evolutionary fantasy, the real focus of life's desire is not other humans but the sun.

Bizarre as it sounds, there is some truth to this. Animal life's activities are not random, but controlled activation of the stored energy of the sun, collected by photosynthetic life, for which sunlight is the source of energy no less than food is ours, and toward which the leaves of plants turn.

In the cosmic scheme of things, it is not sex that is life's true aim so much as reproduction. And reproduction, scientifically viewed, represents the maintenance of a certain kind of complex system that uses available energy, spreading it in the process. Such systems, not confined to life, spread energy and are favored by a universe obeying thermodynamics' second law. The second law, usually still construed as a tendency toward disorder, is really a more general principle easier to visualize as the spreading of energy, which in the chemical realm leads to complex molecules. But in some regions of matter, the path of least resistance is for the continuous construction of complex and ordered structures that more effectively dissipate energy.

Thus tornadoes and other complex chemical systems—and life is nothing if not a complex chemical system—not only arise naturally but, if they've tapped into a reliable energy source, tend to expand until they've made use of all available energy.

As complex systems maintain and grow, differences, such as those between hot and cold, and high and low pressures, chemically concentrated and less concentrated adjacent regions tend naturally to even out. Complex whirling typhoons spreading above the Pacific, for example, reduce the difference, or gradient, between high- and low-air-pressure masses. If it can, nature will produce complex systems, such as convection cells, whirlpools, and repeating chemical reactions, that are better at dispersing energy than mere random arrangements of matter. In these terms, life belongs to a class of complex systems cycling matter to spread energy, finding where it is concentrated, using it and dispersing it, mostly as heat, in accord with the second law. Geophysicist-ecologist Eric D. Schneider points to satellite and airplane measurements above ecosystems that show life spreads energy and reduces gradients like other natural complex systems. Evolution itself is not random, but is naturally oriented to the depletion of energy reserves, the greatest of which is the sun. Stars would burn out anyway, but life on Earth is part of this process. Life measurably keeps itself cool as it dumps heat into space. It is thus no coincidence that life is focused on the sun. Sex only exists because of life, and life as we know it, on the vast scale we know it, only exists because of the colossal local energy concentration it is helping to spread out (to spend, as Bataille would put it) as it eats, grows, and reproduces—the sun.

Bataille was right to emphasize the sun as object of life's evolutionary desire.

The entire energy we see in life—including the transgressive tangents Bataille's teenagers take as they move through a series of unusual love objects ranging from liquids like cat's milk,

semen, urine, blood, and tears to relative solids like eyeballs and eggs—is a displacement, a permutation of the energy of the sun. The great educator and rebel Giordano Bruno was burned at the stake as a heretic in the central Roman square of Campo de' Fiori on February 17, 1600, for, among other things, believing that there were other living worlds besides Earth and that Mary was not a virgin. He also believed, correctly it turns out, in a sort of natural reincarnation where matter recycles to re-form the constituents of all living beings: "Don't you see," he wrote, "that that which was seed will get green herb and herb will turn into ear and ear into bread? Bread will turn into nutrient liquid, which produces blood, from blood semen, embryo, men, corpse, Earth, rock and mineral and thus matter will change its form ever and ever and is capable of taking any natural form . . ."

These material transformations are part of the natural cycling of matter in regions of energy flow. This orderly cycling is only possible because of external energy sources that power the maintenance and growth and, in life, the reproduction of complex systems in accord with the second law. Such cycling transformations come prior to meaning—whether beautiful music or Bataille's theoretical cacophony—and the search for it. That Bataille is considered to be a source for postmodernism is no coincidence. Having read, in Paris, in 1929, *La Biosphère,* the book in which the author, Russian geochemist Vladimir Vernadsky, for the first time regards life as a naturalistic energetic whole transforming solar energy, Bataille's thought was infused by a powerful strain of science that had nothing to do with morality except by its conspicuous absence. Torn later between the competing ideologies of communism and fascism, Bataille anchored his thought in the theoretical bedrock of Vernadsky's energy science. Life's transformation of solar energy is beyond our morality, older than our species, and the source of our erotic energy and sexual obsessions.

Cosmic Love Machine

I knew you when you were a gleam in your father's eye. This joking demarcation of our existence's lustful beginning is trumped, humor-wise, by the variation, I knew you when you were a leer in your father's eye. To have all the advantages life has to offer it is crucial to choose your parents well. Of course, riding the tiger means that we do not get to choose. We are dealt certain cards before we are born, although how we play them may be up to us. It is a source of embarrassment to some to contemplate their beginnings in the base acts of parental coitus.

We say *beginnings* in such descriptions, but that is really a convention, as the life cycle is indeed a cycle. We are thrown into the world like dice tossed by a trickster god. Where were you before you were born? Everywhere and nowhere, I'd wager. Your identity is that part of the cycle illuminated by the beacon of consciousness and the ego attachments of a biological being heading fearfully toward his death. But the process itself goes on. The getting together of sperm and egg to produce a sexually merged cell, a zygote in biological parlance, completes the cycle. If I were writing science fiction, I might have an alien race of angels come to earth and make love to apes, then take off in UFOs, leaving us with cultural memories of heaven. But I have found that the more satisfying a story is, the less likely it is to be true. We are probably not hybrids of departed angels but aspects of an immensely robust and ancient complex system integrated and existing as

the surface of a planet circling around a medium-size yellow star on the outskirts of the flying spiral disk we call the Milky Way Galaxy.

Sex, like the universe, is an energetic phenomenon. Just after the peak of the male's momentary pleasure, to use Chesterfield's phrase, the male ejaculates. Oxytocin, the "cuddle hormone," briefly builds up in the bloodstream. The female, if she reaches climax, and many, especially prior to childbirth, do not, may experience uterine contractions that produce a vacuum in the womb, increasing the chances of the semen being retained and pregnancy ensuing. The rise in oxytocin in her bloodstream compared with his will be orders of magnitude greater, and is one reason that women after mating may feel bonded to the man they had sex with. It takes energy to make babies, and babies require energy to grow.

We tend to mark death as the end of the natural life cycle, birth as its beginning. But the Chinese count one's age not from birth but from conception, which could ensue very shortly after ejaculation. The ejaculate that punctuates our life cycle, opaque and milkily opalescent semen, is 90 percent water—like tears, guts, and the heart. Like most of what lies under the waterproof skin, semen reflects life's aquatic origins. The energy-driven cycling of life retains the environmental conditions of its origins. Life makes use of carbon, hydrogen, oxygen, nitrogen, phosphorus, and sulfur atoms combined into water, sugar, DNA, and other hydrogen-rich compounds.

Humans and other animals aren't the only ejaculators. Cycads and ginkgos—the latter plants are sold in pill form to improve the memory—produce swimming sperm, for example. You won't see their images plastered on video cases in ye olde sex shoppe, but that doesn't stop them from doing the nasty. Unlike most flowering plants, ginkgos are gymnosperms that were around during the Mesozoic times of giant chew-

ing *Apatosaurus* (formerly *Brontosaurus*), *Diplodocus,* and their dinosaurian ilk. The seeds that result from the eggs fertilized by their swimming sperm aren't enclosed in fruit. (*Gymnos* is Greek for "naked"—etymologically connected to *gym* and *gymnasium,* because the Spartans used to take their military exercise naked.)

Each day the female ginkgo can make hundreds of stinky fleshy cones, one seed each and emitting a rotting flesh smell. The odor is from butyric acid, also found in vomit. Some theorize the stench convinced hungry dinosaurs to walk on by. But in Mesozoic times these seeds may also have been dispersed by now extinct animals for whom the butyric acid, like some exotic primeval cheese, was attractive. In any case, the sperm-emitting ginkgos and their partners are still doing their stuff, making naked seeds containing tiny plant embryos that drop en masse to the ground.

The ancestors of both plants and animals were fertilized in watery mediums, but not necessarily in one another's bodies. Animals evolved earlier, but plants came to land first, perhaps with the help of fungi that delivered nutrients to their roots, and animals weren't far behind. The formation of an embryo, a complex growing house of cells that results when an egg is fertilized, is a key defining feature of both plants and animals. Ginkgos and salamanders and chestnut trees all make embryos. And it is sex that gets the ball of the embryo going, its stem cells dividing and then differentiating. The stem cells in the early days of an animal can differentiate into tissues and organs, something plants can do throughout their lives.

In our species it is rare for a person to combine both sex parts, and impossible for the hermaphrodites that do exist to fertilize themselves. But only 5 to 7 percent of plants have males and females separated into distinct individuals, in the way that most animals are. In 20 to 30 percent of plant

species male and female flowers grow on the same organism. And most flowering plants, such as tulips, carry both sexes in the same flower. Nonflowering plants, such as mosses, can produce sperm from leafy tips that swim through water on the surface of the plant; conifers have male and female cones, the male cones with pollen grains that are blown by the wind and fertilize egg cells if they land on a female cone, often on the same plant; or plants may separate into male and female trees. Ginkgos are dioecious, meaning the two sexes tend to be separated into separate bodies, as in us. Wind carries the sperm nuclei in male bodies that alight on the females and then swim toward their goal, the virgin egg.

The oldest ejaculation in the fossil record occurs in the Devonian, 408 to 363 million years ago. The paleological preservation of the salacious scene is reminiscent of the erotic frescoes of erotic paintings preserved at the village of Mount Pompeii by the volcano Vesuvius. Long before humanity, the earliest preserved ejaculation took place among a member of the Rhyniophyta, a phylum that contains the first land plants, which began to diversify some four hundred million years ago. The act was caught *in flagrante delicto* by chert, fine crystalline quartz with an uncanny ability to preserve fossils. *Aglaophyton major* is one of the most common plants in the volcanically preserved Rhynie Chert, named for the nearby village of Rhynie in Aberdeenshire, Scotland. Early in the colonization of land by plants, and before the evolution of true leaves, the fast-setting minerals preserved a host of petrified plant, fungal, lichen, and animal specimens. (Animals evolved earlier but came to land after plants.)

Since sex usually occurs in water, it doesn't tend to preserve well. But in one four-hundred-million-year-old silica-rich deposit local changes in pH remobilized some of the silica, leaving behind thin films of the original organic material. In the specimen the chert beautifully preserves the plant's deli-

cate archegonium (from *goni,* Hindi for "sack," akin to *yoni,* Sanskrit for "vagina")—the female sex organ. Another sample of rock, sliced thin and observed with a microscope, shows *Aglaophyton*'s antheridium, its male sex organ—filled with sperm cells ready to explode. Here, preserved by chance, with neither compromised actors nor moral qualm, is a geographic equivalent of the "money shot" of pornographic films—an ejaculation event 140,000 times older than Homer's *Odyssey,* 400 times older than the human species, and almost as old as the appearance of animals in the fossil record.

Then, as with sperm today, including ginkgo and moss sperm, the sperm tails were constructed of pairs of tiny tubes, usually nine in a circle surrounding a central pair. In modern-day sexual intercourse these tails, still found in many microbes and plants, break off. Each sperm head, without its whip-like tail, called an undulipodium, enters the egg. Like the rest of fabulously varying biology, sperm structure varies from species to species and within species. Some sperm may be slow swimmers, others propelled with double tails. Sperm of the hardy sand-growing cycads are noteworthy. These plants were quite successful during the Jurassic, when they were a favorite dinosaur food. They look like ferns and still grow in the tropics and subtropics. But a single cycad sperm may have forty thousand spiraling sperm tails and be up to a sixtieth of an inch long, big enough to be seen by the naked eye.

Whatever the species, the undulating tail propels the sperm head with its load of nuclear genes toward its destination, the heart of the egg. The problem now is that there are also other cell structures found in the sperm, specifically the mitochondria, the oxygen-metabolizing "powerhouses" of the cell, that will only confuse the cellular apparatus of the egg, which already has her own. Either the male mitochondria don't enter or—if they do come in—they are destroyed, while the genes in the nucleus are welcomed.

The googly eyes of your nattily dressed, sleek-skinned lover—or the jealous glare of your ball and chain—belong to a body, hot or not, that has grown from the merged cell of a zygote, a fertilized egg. The growth has come about by reproduction at the level of the cell, each new cell inheriting the doubled DNA, the extra set of chromosomes—one from each parent—that were dovetailed together in the process of fertilization. The biological name for reproduction in this fused state, allowing the body to grow by making new cells, is mitosis. Mitosis, the "dance of the chromosomes," differs from species to species but its essence is simple: The chromosomes, containing DNA, replicate, then the rest of the cell does, then the cell separates from itself, making two new cells. Mitosis doesn't change the number of chromosomes; it doesn't meddle with the doubled DNA courtesy of the zygotic pairing of Mom and Pop.

Plants and animals can't grow indefinitely in the diploid cellular state of having two sets of chromosomes, however. Like a city, the cell population of the body bumps up against natural limits to growth. Indefinite mitotic cell growth, without regard for the proper borders of the tissues and organs of the beautifully integrated complex genetic machine that is the body, has a scary name: cancer. Once they differentiate, cells can only reproduce so many times. The end of the chromosome (the telomere) is like the tip on the end of shoelaces that keeps them from unraveling. In most cases, each time a chromosome reproduces this tip gets shorter, leading to a limit on the number of times most cells can reproduce. Under normal conditions a cell will only be able to reproduce some fifty to seventy times. (Stem cells produce an enzyme, telomerase, that allows them to break down their telomeres, thus continuing to grow.) Telomeres prevent chromosomes from fusing, rearranging, and continuing to grow indefinitely, which in a way is the normal state of single cells, before they

discovered sex, that is, and growing together in multicell bodies. Unrestrained growth of cells in a differentiated body can spell premature death for the whole system, however. The metastases of cancer turn the body to a more ancient, but less organized state.* Telomeres thus seem to keep cell growth in check by imposing a natural limit on the number of cell divisions. But the price paid for this lack of immortality on the cell level is high. Even though it helps shape the exquisite organization of the animal as it grows from embryo to adult, built-in restraints on cell-level reproduction ultimately lead to death for the organism as a whole. For the beautifully integrated organism to function smoothly, it cannot grow indefinitely, but must go back to the starting point of the cycle, and make a new sex machine.

At the cell level, this requires not mitosis but its sister process, meiosis, the result of which is sperm and eggs. Meiosis starts in the testes in males and the ovaries in females. Meiosis is a rare form of cell reproduction in the human body because instead of keeping the same number of chromosomes, it reduces them by half. After several cell divisions, the offspring of somatic (body) cells—sperm and eggs (also spores in plants)—come together to renew the cycle.

Although the cells of our bodies, except for the sperm and egg sex cells, are diploid—they have two sets of chromosomes—all plants have a life phase (called the gametophyte stage) that we don't have, in which they produce bodies in the haploid state, with each cell having only one set of chromosomes. For example the female "plant genital" archegonium above, which produces egg cells, and its male antheridium

---

* Life, a thermodynamic-energy-driven system, is not unlike other complex systems in this regard. Even nonliving complex systems, such as man-made whirlpools formed in laboratory apparatus from a rotational pressure gradient, reduce their number, vigor, and complexity when deprived of sufficient available energy, or the means to disperse it.

counterpart are organs made of cells each with only one set of chromosomes, so no meiosis is necessary to halve the number of chromosomes. Sperm are made directly via mitotic division. Their swim toward the sweet smell of success is triggered not by short skirts or words of amorous devotion, but by dewdrops, rainfall, or other wetness that carries chemical attractants released by ripe archegonia.

The complex energy-driven life cycles of plants and animals that feature haploid sex cells with diploid bodies permitted new things in evolution. Without the complex sex cycles of embryo-growing beings there would be neither fragrant springtime with its hillside wildflowers nor the promises or poems of the paramour who plucks them. There are other side benefits, like brains. Bones, blood, brains, and brawn are known only from the diploid phase of animal bodies. Without these intriguing meat computers, we would not have the means to grasp our own place in the cosmic energy cycles that use sex to restart embryonic growth.

The sexual reproductive cycles that got swinging, not even a billion years ago, brought with them a frightening complementary motion, the switching from side to side of the Grim Reaper's scythe. With meiosis came mortality because going back to sperm and eggs eventually meant discarding those trillions of somatic cells that, although having brilliantly served their purpose, were not directly represented in evolution. And with reproductive sex came programmed cell and differentiated body death, because evolutionarily our bodies are husks, biodegradable reserves of valuable bioelements that belong to the ecosystem and must be returned, like overdue books, after performing their natural duty of keeping going the larger energetic process. Personally, as intelligent animals, we identify as individual bodies. Although easier said than done, the mystics advocate a larger view in which we identify with the cycles of natural energy-transforming forms, as well as release from

such life cycles, which they call nirvana. Creeping behind the bright prospect of Mesozoic ginkgo-sniffing reptiles, primeval ejaculators, and the first fragrant flowers was that dark figure, the inevitability of their demise. A melancholy note was struck in the cosmic love machine.

Seduced by Sade

*T*he Marquis de Sade, from whom we get the designa-
tion *sadist*—not to be confused with Sacher-Masoch,
from whom we get the tag *masochist,* although, as we saw in
chapter 5, he had peculiar tastes, too—was one twisted individual. No one in the history of literature has depicted so
many bizarre and cruel sex acts between the pages of so many
books, many of them burned, including by his own family.
And yet there is something strangely honest about his life, and
curiously consistent about his philosophy. His was not the path
of denials and hypocrisy, repressions and secrets. The paper
bodies spouting sexual philosophy and engaging in elaborate
orgies in his fiction were used for means beyond the mere
pornographic titillation of readers. They had an educational
end: to show that nature is what it is, and no amount of moral-
izing hypocritical cant from human beings is going to stop it.

Initially imprisoned for excesses committed in a brothel that
he frequented for a month after marrying the daughter of a
military official, Sade talked his way out of imprisonment by
saying he thought such behavior was normal for aristocrats.
Four years later he got in trouble again for abusing a beggar
woman, Rose Kellar, taking her to one of his residences in
the southern suburbs of Paris and dripping candle wax on her
naked body, among other things. He served four months. Then
in 1772 he was condemned to death for giving Spanish-fly-
laden aniseed candies to prostitutes, including one who developed severe stomach problems. Spanish fly is an emerald-green

beetle, 5 percent of whose body weight is cantharidin, a urogenital stimulant given to farm animals to get them to mate. It can cause erections in small doses and permanent kidney damage in large doses. By the time the death sentence had been handed down, Sade had already fled to Italy from his castle with his sister-in-law, Lady Anne, with whom he was having an affair. Not being around to enjoy their punishment, both Sade and Latour, his manservant and the traveling companion who had procured the young prostitutues, aged between eighteen and twenty-three, were executed in effigy on the Place des Precheurs, in Aix en Provence on September 12, 1772.

On the run, Sade under the assumed name of Count de Mazan, with Latour and Lady Anne, leaves his luggage in Nice and holes up in Chambery before the law (alerted by his mother-in-law) catches up with him and he is imprisoned in Fort Miolans. As will be typical throughout his incarcerations, Sade complains about the frightful condition of the prison (when he is not attempting to seduce prisoners). His wife writes to the king of Sardinia that, "My husband is not to be classed with the rogues of whom the universe should be purged . . . Bias against him has turned [a misdemeanor] into a crime," whereas his was a "youthful folly that endangered no life nor honor nor the reputation of any citizen . . ." After four months he escapes with another prisoner and, still sentenced to death by the High Court of Provence, makes his way back to his castle (whose ruins in recent days have been refurbished by French clothier Pierre Cardin) in Lacoste in Provence.

In 1778, after doing time in Vincennes for abusing servants—and with lobbying this time on his behalf by his mother-in-law (who also told the authorities where to find him)—Sade beats the poisoning charges stemming from the Marseilles Spanish fly incident, which are dismissed for lack of evidence. The focus on his deviancy leads to a new trial where he is found guilty of acts of debauchery and excessive libertinage, fined

fifty livres, and ordered to stay out of Marseilles for three years. However, he is still sentenced to Vincennes for 1777 charges brought against him by the pistol-wielding father of one of the servants, Catherine Trillet, who is very pretty and known in the castle as Justine. (A monk, a friend of Sade, assured Trillet's father before he sent his daughter there that the conditions of the castle are like "a nunnery.") Evidence of the marquis's evil charm, three days after shooting point-blank at Sade (but missing), Monsieur Trillet, the beautiful young woman's would-be homicidal father, tells mutual acquaintances of his "sincerest emotions of friendship and attachment for the Marquis."

On the way back to Vincennes, Sade briefly escapes but is recaptured and transferred in 1784 to the Bastille, the famous tower that served as a prison in central Paris. Several days before the Bastille was stormed—inaugurating the 1789 French Revolution—Sade is heard yelling down to the street from one of the windows that they are killing the prisoners in there and the people should liberate them. Two days later, as a result of a report on this behavior, Sade is sent to Charenton, the insane asylum, from which he is released when, after the bloody revolution, the Constituent Assembly invalidates Royal Orders, except for those condemned to death, indicted, or formally judged insane, none of which criteria applies to Sade. He emerges broke but adjusts to the social upheaval of postrevolution France, entering politics and even briefly becoming a magistrate before his reputation and continued publishing of obscene works catches up with him. In 1801 he receives "administrative punishment" for being author of "that infamous novel *Justine*" and the "still more terrible work *Juliette*" and is sent to Sainte-Pélagie, a former convent become political prison. Denying authorship and complaining of conditions, he is sent to Bicêtre prison and, continuing to complain, is transferred to back to Charenton after his family agrees to pay his bills. He spends his last years confined to the

insane asylum, directing plays and winding up his four-year affair with Madeleine Leclerc, who is seventeen when Sade, obese, dies in 1814 at age seventy-two.

Sade was called the "freest spirit that has yet existed" by French poet Appolinaire (who coined the term *surrealism* and who was himself arrested in 1911, under suspicion for stealing the *Mona Lisa,* though later released), though the extent of overlap between Sade's life and his work is unclear. Philosopher Pierre Klossowski has argued that Sade was railing against the meaninglessness of a godless world where man's power to overcome nature was ultimately in vain—nature would always triumph in the end. In his writings, the bisexual aristocrat seemed in any case determined to show the folly of Christian virtues of chastity, goodness, and temperance. Man was not in control of anything; really it was nature that would have her way and could not be resisted. In *Philosophy in the Bedroom,* one of Sade's speech-making characters preaches the pleasures of anal intercourse, pontificating on how useful it would be if the lower orders practiced it en masse, as the devoutly-to-be-wished-for result would be the extinction of the human species. It is hard to take such a comment, delivered sanctimoniously by an orgiast, entirely seriously. Were the preachings of this most famous pervert in history to be taken seriously? Did he himself know?

Surprisingly, the great feminist writer Simone de Beauvoir called Sade "a great moralist." What she had in mind is also difficult to say, unless she was implying that the evil portrayed in the Marquis's vast pornographic corpus was tinged with an irony as profound as it was subtle. Or maybe she had in mind the exact opposite: that his "morality," unlike the socially acceptable one preached by the consensus, contained not the faintest whiff of hypocrisy. Or maybe somehow she meant both. In any case, his writing was an authentic attempt to think about freedom and engage in a politics of rebellion, she felt,

but he failed as an artist and was unable to develop an ethics of genuine emotion beyond his staged cruelty. (Sade claimed later in life that he had never done nor would he ever do most of the things he wrote about.)

Beauvoir pointed out that to abuse the flesh you have to value it, implying that traditional moralists did not. It might be said that Sade's "morality" was a caricature of confused human ideas of absolute right and wrong, exaggerating for the sake of clarity the struggle to overcome ambiguity when nature herself is immoral. The confusion was redoubled by the 1789 overthrow of the privileged classes, the noblemen and royalty whose superiority supposedly derived from God and king, and by the subsequent beheading of the king and queen. Sade's exhaustive platitudes showed that the king of morality was without clothes, that there was no absolute moral force ensuring good behavior. His own perversions commingled with the crumbling authority of a society crying out for freedom but losing its moral bearings. If the American Revolution in 1776 showed that new societies, free of monarchy and colonial control, were possible, the French Revolution in 1789 showed that overthrowing rigid class structures could lead to a chaos more frightening than the caricatures of rationality-steeped sexual proselytizers, the depraved aristocrats of Sade's imagination.

Nature as we know it, like other complex systems, recycles matter as it maintains and, where possible, expands its ordered realm, making use of energy to do so. It is for this reason that organisms devour one another and evolve ways to counter predators, be these bodily armor, speed in running, or intelligence itself. Through the physical system energy flows, genes spread, and matter cycles. Establishing a moral code to guide us through this process is fraught with difficulty. Doing so may excite our noble sentiments and satisfy our enthusiasm to recognize emotional attachment and mammalian love. But rationally speaking, the arbitrary and the hypocritical lie

in wait at every turn. Like the Monty Python song "Every Sperm Is Sacred," noble sentiments to preserve all life-forms, no matter how small, in the end are ludicrous. We might try people for murder for killing millions of microbes each time they use a Handi-Wipe or spray Lysol on their kitchen counter. But talking the moral talk is not walking the moral walk. "All universal moral principles are idle fancies," wrote Sade: "All, all is theft, all is unceasing and rigorous competition in nature; the desire to make off with the substance of others is the foremost—the most legitimate—passion nature has bred into us and, without doubt, the most agreeable one."

Not only was Sade painfully aware of the waning privileges he would enjoy as a nobleman, but he lived to jump ship and proclaim himself a commoner as the nobles were stripped of their power. Compared with the bloody aftermath of the storming of the Bastille, the crimes for which Sade served so much time (not including a stint in debtor prison) were considered trivial by the early 1790s, when French society, having undergone a revolution, was now convulsed by "the Terror"—in which some forty thousand people were put to death without trial. Part of the rage directed at the upper classes was for its extreme taste for its own *plaisir*—pleasure—and here Sade was not unique but emblematic of the aristocracy's cavalier use of others to satisfy their sexual needs and elaborate lifestyles. King Louis XV's erotic affairs and illegitimate children were said to be a strain on the national finances, contributing to the economic climate that made a revolution possible—and fateful for Europe because, for the first time, royalty who claimed their absolute privilege came from God were violently deposed—a climate of steadfast rebellion that made places like the United States, which France helped finance (adding to its financial turmoil), possible.

Only fourteen years after the start of the American Revolution, for which France had supplied resources (including officers and ships), Sade was elected to the National Convention.

He called himself Citizen Sade and lobbied for the direct vote. Despite the numbingly repeated depictions of sexual violence in his fiction, in real life Sade abhorred the idea of capital punishment. The brutality of real politics upset him. Horrified by the Terror's excesses under Robespierre, Sade resigned from politics in 1793. In one of history's unwitting ironies, he was imprisoned for a year (in Madelonnettes) on charges of "moderatism."

Certainly no one would accuse the man, whose valet Latour had, among his duties, that of sodomizing his master, to be the incarnation of propriety, a poster boy for the respectful keeping of boundaries. But the point is he had a point, however confusing, or confusingly made: His was a rationality of sexuality that embraced the Enlightenment's ideal of reason while refusing to disavow sexuality or the conflicting feelings that attended it. "It is not my [logical] mode of thought that has caused my misfortunes, but the mode of thought of others." He viewed himself not as the source but as an honest broker for "lust's passion" that demands to be "served . . . [and that] militates . . . tyrannizes." His antipathy toward state hypocrisy can be seen in some of his simple but sensible political opinions, for example that the state had no right to make murder a crime with one hand while it enforced the death penalty and sent men to war with the other. "Nature" speaks not with "two voices, you know, one of them condemning all day what the other commands," but with a singleness of purpose, and it is as ridiculous to restrain our sexual appetites as it is our desire to eat or drink. Nature suffices without God, war is the highest immorality, and freedom demands that we compensate ourselves in "privacy for that cruel chastity we are obliged to display in public." In short, Sade, the "evil Marquis," for all his crimes and failings, was at heart honest in his insistence that humanity face truthfully nature's unseemly nature—and admit to being part of it.

By satisfying his seemingly inexhaustible sexual appetites as much as he could in life, and more so in fiction, he mocked man's moral hypocrisy and his attempts to overcome nature, even as he attempted to preserve his aristocratic privileges and extend his sexual freedoms. "Nature," says Katharine Hepburn (who in real life braved dysentery, malaria, and crocodiles in the making of John Huston's *The African Queen*) to a gin-swilling Humphrey Bogart, "is what we were put on Earth to rise above."

Far from rising above it, however, Sade wallows in its depths, lecturing us all the while. When he stages rape and mayhem, the most twisted of scenes, he is simultaneously vetting his diminishing powers as a member of a socially privileged class, and scoffing at the notion of an infinite power with a moral compass. If the king, earthly enactor of the divine, can be killed, where does that leave God? When Sade fictionally dreams, as he did at the end of *La Philosophie dans le Boudoir,* of sewing rapists' semen up in the vaginas of the proper wife and the mother who would protect and save her for a proper marriage, he marries the cruelty of a criminal mind with the pedantry of a philosopher who will take no prisoners, make no compromises in facing the blinding sun of natural process. He writes as if to say, Here is amoral nature, with no moral compass whatsoever, exaggerated by me for your viewing pleasure. This is what you moralists would put forth as a model if you preached what you practiced, if your teachings were not papered over with social hypocrisy and habitual self-deception. Behind the perversion is a monomaniacal sense of purpose, itself defiant, to display nature's amoral sexual working materials with philosophical clarity and nonhypocritical consistency. This is what Beauvoir meant when she said he was a great moralist. And why she thought he came up empty-handed in his analysis of ethics.

If the consistency Sade displayed was only the ad hoc ranting of an oversexed sociopath, it would not fascinate us so

much. But the consistency he displays, the logic of his lust if not his proselytizing zeal, is familiar to us because it goes much deeper. And it is not that of the devil. The consistency Sade displays through the characters of his books and, to a lesser extent, in his own life, is that of an amoral, matter-cycling, energy-using, energy-attracted complex system. We are mammals with strong emotions, but the complex system doesn't care about our feelings. In the long run, it doesn't even care about the human species that, Sade fantasizes with sadism but romantic irony as well, would vanish if the entire populace were to recognize the superior pleasure afforded by anal over vaginal intercourse. This fantasy, the ultimate perversion, is Sade's way of claiming his freedom against the unfeeling determinism of a Godless universe. If you can't beat them, join them. Sade, so to speak, did both.

When Sade claims there is no absolute morality, he is voicing an uncomfortable truth recognizable to scientists and religionists alike. Discussing the God-given rights of man and how they differ in number and content from nation to nation, comedian George Carlin has wondered whether we are supposed to assume that God, of all entities, is bad at math. And whether or not rules, like records, are made to be broken, these various moral prescriptions are much easier declared than adhered to. All you have to do is watch *American Idol*, read the divorce statistics (lower among atheists), notice the peccadilloes of self-appointed moral leaders, or peruse newspaper accounts of wars to realize that the most famous moral prescription in history, that of Moses on the mount, is adhered to quite spottily. In fact it seems safe to say that moral prescriptions are concocted precisely because it has become painfully obvious that the prescribed behaviors in question are most assuredly *not* being adhered to. If they were, there would be no reason for the prominent mention.

Realizing that religious moral codes have no natural basis, some scientists and ethicists have looked to science for guidance on how to do the right thing. Unfortunately, it is extremely easy—as Sade so ably shows—for tyrants, exploitative businessmen, and sociopaths to find shining precedents and wonderful guidelines for whatever appalling thing they have in mind. Despicable criminals are wusses compared with the life cycles of some insects. African bedbug males, for example, routinely pierce females through any part of their carapace in order to impregnate them. As nature is no more moral than Sade, the excitable males also pierce one another. Because of their circulatory system, which keeps the sperm alive, the pierced males may then, bizarrely, impregnate the females with whom *they* mate with the sperm of the males that penetrated them—natural sadism in action. Natural selection is amoral. The kinkiest humans are utterly unimaginative next to the sexual variety on display in the animal world. Which dysfunctional human family is comparable to the behavior of *Adactylidium,* the nasal mites whose mothers give rise to 99 percent females? Inside the mite's womb, one unborn male inseminates all his unborn sisters, then dies. The incestuously impregnated females grow, eating their mother's body from the inside out to provide themselves with nourishment. They are born pregnant, beginning the cycle again.

Some insist, nonetheless, that we can look to nature for moral guidance. Philosopher Fred Turner, for example, argues that our emotions regarding moral behavior are not arbitrary but have evolved, and that they should be consulted for clues in making sensible, nature-based moral rules. For example, and provocatively, Turner argues that it is useless to try to root revenge out of human morality. Vengeance is a strong and perfectly natural human emotion and serves a punitive function central to jurisprudence. Legislating against it would be doomed to failure. In Sade's case, he worked out his desires

for vengeance against his mother-in-law fictionally in works like *Philosophie dans le Boudoir,* and this perhaps helped him resist doing so in real life when he, as a citizen officer after the revolution, had a chance to wreak vengeance upon her. Even though he had served two prison sentences because of her, he instead let the matter drop, moving the Montreuils to a list of exculpated persons.

But this is a tough row to hoe. Even if legislators and would-be moralists could be convinced of the scientific superiority of looking for guidance to the activities of nature, human nature in particular and evolutionary precedent in general, they wouldn't agree on the details. Some would warn of social Darwinism, the use of evolutionary theory to excuse abusing workers, employing children for labor, and so on, in the name of survival of the fittest. Worse, the Nazis used an appeal to nature, their questionable interpretation of the Nietzschean philosophy of the strong will and the overcoming of weak modern man, to try to engineer a new superspecies by eutha-nasia, sterilization, and genocide. Discovering that traditional morality has no natural basis does not mean that nature can provide one.

More hopeful than coming up with a universal moral code is the less grandiose task of cultivating interpersonal respect and a local ethics. From a religious viewpoint, the Jewish philos-opher Emmanuel Levinas argues for a switch from what he calls prescriptive ethics—the Ten Commandments and similar moral codes—to descriptive ethics: being responsible in front of the face of the other. Immanuel Kant spoke of the cate-gorical imperative: "Treat others as ends, not means." Sade's failing was that he systematically treated others as means, not ends. They were there to be used, either as objects of plea-sure, or as mouthpieces for his philosophy—which was itself to treat others as objects, because that's what, by his lights, nature does.

Simone de Beauvoir argued that in the staging of his philosophical points, even in fiction, Sade ignores the ethics of the other, treating people as extreme (sex) objects rather than feeling beings with their own rights and desires. The middle ground between a prescriptive and descriptive ethics is the golden rule, associated with Jesus but espoused earlier by Confucius and others: Do unto others as you would have them do unto you. (Jesus may have taken it from his elder, Rabbi Hillel [circa 30 BCE–AD 10], whose most famous maxim was, "Do not unto others that which is hateful unto thee.") This of course implies the opposite: Don't do unto others as you would not have them do unto you. Megalomaniacs, sociopaths, and politicians have some difficulty with this concept. And some confuse it with what might be called the Machiavellian Rule: Do unto others before they do unto you. In his life, however, Sade seems to have adhered to a version of the Golden Rule, being keen on taking it as well as dishing it out. It is true that he violated the categorical imperative, treating others as a means—for sexual pleasure—rather than as ends in themselves. Which goes to the heart of the feminist complaint about women being treated as sex objects.

Spiritually and sensually, sex encompasses far more than pleasure or reproduction. One day at lunch Hitler, a devotee of Nietzschean philosophy (who conveniently ignored Nietzsche's disparaging remarks about his German countryman), remarked, "As in everything, nature is the best instructor." But nature's instruction is not very clear. Human nature is part of nature and it is our nature to question nature, and to question in general, which is at the heart of both philosophy and science. To blindly follow a mad and contradictory master is a road leading to the madhouse, not enlightenment. The former nun Karen Armstrong in her detailed comparison of world religions argues that the one thing they have in common is the golden rule, to not treat others in ways you

would not have them treat you. This spiritual kernel shared by Christianity, Islam, and Buddhism, among others, is an argument for consensuality, not coercion. Sex represents the height of human intimacy and, as Christopher Isherwood points out, no friendship is greater than that of former lovers who remain close.

Nonetheless, Sade practiced his own twisted version of the Golden Rule, only it was with nature not man. If he could dish it out sexually, he could also take it, and if he could not beat nature he would join her. Moreover, in the last analysis, his fictional characters, like the celluloid heroes in the Kinks' song, never felt any pain. Unreal, they never suffered, even those in the flames of the final unpublished volumes, written at Charenton, burned before reading by his eldest son. If he subjected them to endless abuses, treating them as means to a dubious educational end, he wished for himself a curiously consistent fate, that he be buried in an unmarked grave, allowed to grow wild, so that "the traces of my resting place shall disappear from the face of the earth as I flatter myself that my memory will be erased from the minds of men."

Laugh of the Hyena

*T*he story goes that when the politician, playwright, and substantial wit Edward George Earle Lytton Bulwer-Lytton, First Baron Lytton (1803–1873), a popular novelist in his time, who coined still-famous phrases such as *the great unwashed, the almighty dollar,* and *the pen is mightier than the sword,* was informed by a society woman of delicate sensibility at a dinner party, upon being offered a plate of tongue, that, oh no, she could never eat anything that came from the mouth of a cow, he picked up a bowl and, setting it before her, said: "Here, have an egg."

Today Bulwer-Lytton is remembered mostly for sentences not entirely unlike the preceding long and winding one, celebrated in the comedic contest that bears his name. The inspirational model for horribly convoluted, unintentionally funny sentences is this doozy: *"It was a dark and stormy night; the rain fell in torrents—except at occasional intervals, when it was checked by a violent gust of wind which swept up the streets (for it is in London that our scene lies), rattling along the housetops, and fiercely agitating the scanty flame of the lamps that struggled against the darkness."* So opens Bulwer-Lytton's 1830 novel *Paul Clifford.* Apart from everything else, it appears not to have occurred to the former British secretary of state for the colonies that nights tend to be dark. The contest rewards similarly grandiose first lines to imaginary novels. A recent tie for third place, in the category of Romance, is this offering from Bruce Portzer of Seattle, Washington: "Carmen's

romance with Broderick had thus far been like a train ride, not the kind that slowly leaves the station, builds momentum, and then races across the countryside at breathtaking speed, but rather the one that spends all day moving freight cars around at the local steel mill."

Why mention what the *Boise Weekly* calls "the only literary contest that matters"? Well, because it's funny. But also because mannered language presents us with a nice gilded frame to look squarely at our animal nature. Epigrams are an antidote for circumlocutions, straight talk a cure for ovine ordure. Diogenes the Cynic, when asked why he begged from statues, remarked that he was "practicing disappointment." Sex is too often disappointing. Former president LBJ said that there is nothing so overrated as a lousy lay, and nothing so underrated as a good crap. Good sex, like an enjoyable life, is likely to have humor in it, an element of the playful. The realm of thought allows us to step back, to not take things too seriously. The difference between comedy and tragedy, it has been said, is that life is a comedy to those who think and a tragedy to those who feel. As Mel Brooks explained, "Tragedy is when I get a hangnail; comedy is when you fall in a manhole."

It is easier to be sanguine about the problems of another. It is harder for us to regard ourselves with the same objectivity. People take life personally. That is how we are built. But truth and humor tag-team to bring humanity's glorified image of itself down to earth. Not to humiliate but, like satirical Swift, to put us in cosmic context.

That humans are animals—eating, excreting, sexually reproducing animals—cannot be denied. If nature is what we were put on Earth to rise above, as Katharine Hepburn says on the *African Queen,* we have not risen much. And even when our astronauts moonrocket they bring with them their gut microbes, the gene-trading bacilli and cocci, the spirilla and vibrios and roundworms and *Candida albicans* and other

natural inhabitants of our bodies, some of whom were already having orgies as Armstrong spoke of a small step for mankind.

When we act high and mighty, we act mighty high. Perhaps as high as the Yahoos—the imaginary primates in Jonathan Swift's 1726 *Gulliver's Travels*. The Yahoos dig in the mud for pretty stones, at least when they are not shitting on passersby from a great—at least to them—height. Lemuel Gulliver would rather spend time with the calm and rational horses known onomatopoeiacally as the Houyhnhnms. Selfish and mean, people are to be avoided. Compared with us, other mammals seem good and noble, and living sustainably within their ecological means.

Well, the ecological part is true, but there are other mammals that give us a run for our money in mean. Take the spotted, or laughing hyena, a fey predator that inhabits grasslands and semi-arid deserts. The Marquis de Sade called his mother-in-law, Madame Montreuil, "the Hyena." He also, separately, imagined a female character with a clitoris big enough to penetrate her lovers. This gave her a pleasure evolution neglected, direct clitoral stimulation during normal intercourse (normal for Sade that is).

The masculinized woman is a product of Sade's imagination. Not so the hyena, which emerges into the world through a birth canal—which runs the length of her clitoris—at least as twisted. You've heard the term *alpha male* to refer to bossy, powerful, hierarchically dominant males. In human beings they tend to be bigger, richer, both physically stronger and more successful at bedding women than the average male. Indeed, men who win fights experience naturally higher blood levels of testosterone. The testosterone boost increases their confidence and their physical strength. In the unfair world the rich get richer and the cocky get cockier.

Aristotle and Hemingway thought spotted hyenas were hermaphrodites. They're not, though; they're just serious alpha

females, *Crocuta crocuta,* the laughing hyena. Neither in the cat family nor the dog family, they act more like dogs but are evolutionarily closer to cats. They hunt in packs, taking wildebeest and zebras down and devouring them in fifteen minutes flat. They eat the bones. Hyena jaws easily support these animals' entire weight. Together they clear their plate and later, post-digestion, their scat is white, due to the undigested crushed bones. Female packs will drive a female lion away from her kill, although a male lion, the king of the beasts, will intimidate them to move away from their fresh kill. Unlike the more familiar lions, hyena females are larger than the males. The girls hunt in packs while the males, whose social status derives from their mothers, hunt on their own. Their demented-sounding vocalizations give them their common name.

Aristotle and Hemingway were confused because the hyenas seem to have penises. Both the males and the females. But the females don't. They have giant clitorises.

Even in the history of natural history, our sexual expectations are based not only on the overly limited experience of our species, but on our wild imaginations. Aristotle, Plato's knowledge prodigy, theorized that the man's seed turned into a boy if it was hot when it went to a woman; cooler sperm, with less energy, never made it all the way to the mold, becoming a little girl. Plato wrote down dialogues of Socrates, who didn't write but is recorded in *The Symposium* as a love interest of the handsome Alcibiades, a younger man and military aristocrat who was surprised that the philosopher didn't want him. The wings of the angel of love cut like teeth through gums—Platonic love, that is, erotic love without sex, love sublated and hypostatized, elevated from the common realm of flesh to the subtle ethereal realm of thought.

What does it mean to be a real man? Culturally it is questionable. Latin men in prison consider only the receptive homosexual male to be gay. Postmodern theorist Michel Foucault

argued that sexuality itself is a modern conception in the sense that religious confession, psychotherapy, and other recent institutions have led us to identify our sex acts as having a specific orientation and character. The Greeks had slaves, women, and children, all of whom they sexually used. Although homosexual relations were institutionalized in the Spartan military, for example, this did not make their practitioners "homosexual" in the modern sense of having a specific orientation toward the male gender—much as being married, as a matter of family arrangement, in the old days didn't necessarily entail romantic love. Psychoanalyst Jacques Lacan says manhood is a kind of social sham, a "masquerade." Hemingway's mother dressed him as a girl before he became the most macho writer ever.

In biology, the essence of manhood is chemical.* It is testosterone, found in adult men but also, in smaller amounts, in women. The male hormones testosterone and dihydrotestosterone are needed to develop characteristic male features such as the penis, male urethra, prostate, and seminal vesicles. Testosterone in the mother's circulatory system masculinizes the unborn child, fusing the labioscrotal folds to become the midline of the scrotum rather than the opening of the vagina, and turning the genital tubercle, visible after four weeks, into a recognizable penis after nine weeks. In utero testosterone concentration also affects relative finger growth. The longer your fourth finger relative to your index finger (the lower the so-called 2D:4D ratio), the greater was your prenatal exposure to the hormone. In other words, human fetuses exposed to high levels of testosterone grow up to have ring fingers longer than their index fingers. A January 13, 2009, *Proceedings of the National Academy of Sciences* article measured the 2D:4D—the

---

* Although it looms large in our psyches, biologically gender or mating type can boil down to minimal chemical differences: There are ciliates that change mating types due to the presence of a single OH group, that is, one hydrogen and one oxygen atom.

second-to-fourth-digit-length—ratio of London financial traders. (Males typically have 2D:4D ratios lower than one; females,
higher.) Carefully controlled to make sure there were no obvious advantages, the traders who were exposed to higher prenatal testosterone levels as measured by their fingers were six
times more successful in trading. Human prenatal testosterone exposure between nine and eighteen weeks of gestation
sensitizes the body to testosterone's effects later in life. Confidence, vigilance, and a risk appetite are results of increased
testosterone. Multiple studies have found that a low 2D:4D
ratio correlates with athletic skills, from soccer to skiing, as
well as visuospatial ability and protection against heart attack.
The variance in prenatal androgens fluctuating in the mother
depends on the Y chromosome in the unborn male, whose
testes are producing androgens by the eighth week of gestation. Maternal stress can add to the uterine testosterone levels
that produce penis and scrotum rather than clitoris and labia
and which shape nonsexual factors such as finger length and
the brain via androgen receptors there. Some animals produce
more males and more masculinized females when populations
reach a high density. This may account in part for reports of a
higher ratio of boys being born after wars, during famines and
stress. In addition, studies from the 1980s show that dominant
males and female animals tend to produce statistically higher
numbers of male offspring.

Because of its role in muscle development and confidence,
steroids have naturally drawn the attention, and sometimes
elicited the abuse, of competitive professional athletes. Key
word: *competitive*. Steroid masculinizing compounds not only
make males tougher and more anger-prone, but turn females
into queens—at least in the naked mole rats, burrowing
subterranean rodents of such monstrous would-be cuteness
that they are known as "sabertooth sausages." Not all of the
individuals in a population reproduce, only males with suffi-

cient testosterone and the huge queen, whose size and dominance results from the hormone, and whose death triggers replacement by a new female able to reproduce.

But enough about rodents and humans. Not until twenty hyena cubs were grabbed in the wilds of Africa in the 1980s, and brought back to Berkeley, California, did scientists figure out what was going on. The females clitorises were longer (but not thicker) than the penises for the same reason that the females themselves were bigger and stronger—because of steroid hormones. Androstenedione, secreted by the ovaries, is converted by the placenta to testosterone, which masculinizes the fetus. Even after thirty-six days, before its own ovaries can be seen, the formation of a giant clitoris, enclosing the entire urogenital tract, can be observed in utero. It is this giant masculinized clitoris that makes it so hard to tell females from males. But more is going on than mere masculinization, as the entire genital architecture is distinct. As soon as they are a few months old, males and females display full erections upon meeting, in which they sniff and taste each other's genitals. Contrary to expectations, hyena convention requires the subordinate clan member to be the first to submit his or her erection for inspection. Copulation occurs through the elastic end of the tip or glans, the meatus. Males must "flip" their semi-erection in order to search for the opening to the clitoris, which doubles as a vagina and is placed abdominally away from the anus. Before the male mates this highly elastic organ, through which the female also urinates, must be retracted inside of her as if inverting a sock.

Assuming this difficult mating process goes well, an egg is fertilized and grows. At birth, the 3.3-pound embryo must make its way along a treacherous U-shaped birth canal, tearing open the clitoris to arrive alive in the extrauterine world. Which much of the time it does not: 60 percent of first-born hyenas are miscarriages, stillborns that fail to make it to the

starting line. The umbilical cord is only six inches long while the birth canal through the clitoris is a foot. This leads to strangulation. After all this, if they make it, and most of them are twins, they must deal with the sharp incisors and canines of their siblings, which begin to nip and bite. Hyena mothers give birth at the entrance of abandoned aardvark burrows, whose increasingly narrow passageways away from the entrance protects cubs from lions and adult hyenas. The more aggressive hyena cub won't kill her twin, at least not directly. Rather, the constant chomping behavior intimidates the weaker twin. The melancholy result is a retreat deep into the bowels of the narrow set of inaccessible aardvark passageways. Sometimes the weak twin never emerges. She simply stays there, a subordinated baby hyena in an abandoned burrow, when her mother calls. She (or he) is thus deprived of the finer things in hyena life, like frolicking in the communal den, or being the first allowed, as the offspring of an alpha female, to rummage through the flesh of a felled wildebeest.

Biology boasts other tough females. Take the bony fishes, or anglers. These are scary-looking deep-sea fish with an extension on the head. A natural glow-in-the-dark fishing lure, lighting up the ocean abyss, attracts both food and mates. Would-be nibblers of the glowing bait, called the esca, are victims of the proverbial bait-and-switch, being eaten by anglers just as they try to eat. Because of their distendible jaws, flexible bones, and expandable stomachs, some anglers—there are two hundred species of the order Lophiiformes—can swallow and digest fish twice their size.

The natural bait organ glows because of bioluminscent bacteria. Far from being a disease, the symbiotic bacteria are integrated into the angler's body, literally lighting up its life. The bacteria not only act as a deep-sea beacon, but are an example of a permanent, sex-like association between partners from not just different sexes but different species. And

the angler's light in the abyss attracts love as well as lunch: angler males. They are so small, scientists first thought they were parasites of a different species. They got the parasite part right. The angler male, having found the light of his life, sinks his teeth into her body. Thus hooked up, his sharp small teeth lustfully lodged, he never leaves her. More romantically, he produces an enzyme that allows their circulatory systems to merge. Prior to ovulation, her hormone triggers release of his sperm. The hustler's body, fed by her bloodstream, deteriorates. This doesn't stop her from forming new attachments to other mini males as she swims the deep, still a light in the darkness.

In Charles Bukowski's story "Six Inches," a man lucks out with a beautiful woman and things go swimmingly until one day he notices that his clothes seem larger. His troubles increase. He loses weight until his shoes grow as big as a bathtub. He drinks beer by the thimble. Six inches high, he is clutched by the frisky witch, who plunges him headlong into the netherland of her desire, forcing him to time his breaths, so that he may not suffocate of her untoward ardor. Finally, she relaxes her grip. He scampers to find a weapon. Wielding a steak knife like a giant sword, he pays her back, then steals cat food while avoiding dangerously large cats until he regains his stature.

Hierarchical sex relations are very old. It's more than a battle of the sexes, though. Theoretically, female genes may survive better if mixed in children or grandchildren, with male genes conferring testosterone-related bad-boy traits such as violent charm. "The Hyena," that is, the Marquis de Sade's mother-in-law, Madame Montreuil, who spent much of her life trying to get Sade locked up, theoretically benefits—or rather the survival of her genes do—if they are mixed with his in the womanizing male children or grandchildren born of his union with a Madame Montreuil daughter. On the surface she is the

outraged woman of society, morally correct in her derision and justified in her appeal to the authorities. But by turning a blind eye to the marquis's bad behavior, she may initially if unconsciously have been calculating the chances of the future spread of her own selfish genes through children her daughter might have with Sade, and through subsequent generations.

Philosopher Arthur Schopenhauer traces the obsessive human interest in mating relations to a meditation on the composition of the forthcoming generation. Writing before the discovery of genes, he instead personifies the power that attracts people as the "genius of the species," who meditates on "the individual which may be born and the combination of its qualities; and the greatness of [the lovers'] delight in and longing for each other is determined by this meditation." This Cupid-like power makes us weigh one another's traits, and act with a view to mating even sometimes at considerable cost to our own welfare, as what survives in the long run is not us individually but the evolving system of which we are part. A good girl mating a bad boy may have sons who sire more grandchildren for her than if she had stayed with a better man, no matter how moral, faithful, or rich.

The "Hyena," who for years tolerated her son-in-law's infidelities, and then for years relentlessly attempted to restrain and imprison him, must have had deeply ambivalent feelings about a charming user like Donatien Alphonse François de Sade, the evil marquis. Hyena mothers probably don't think twice. Sometimes they put their twin cubs into separate aardvark tunnels to keep the young from attacking each other. Alas, usually these moms are not so considerate, as if preferring to let their cubs battle to the death for the "genius of the species," that their survivors may stand a better chance of life upon the African plains.

Endless Desire

DNA Dance

*S*ex means gender, pleasure, mixture, even genitals, as when Anaïs Nin writes, ". . . her tongue . . . the tip of his sex."

In evolutionary terms, though, sex always comes back to the genetic dance of recombination. The chromosomes join forces in the same cell. They even line up next to one another and switch segments. Geneticists call this crossing over. The mixing is extensive.

But why? Why all the shuffling? Can't organisms reproduce more quickly without it? Many phyla in the animal kingdom have representatives that can reproduce with only one parent. Why should a powerful woman like Madonna have to bother with husbands or trainers when she could have a baby without heartbreak just by budding? Well, not Madonna, of course, because the technology is not available. But if so many organisms can get away with skipping the part where you have to mate with a male, and it is more efficient, why haven't they all evolved like this? Why see men and why semen, to raise a human version of the "damnable expense" question I raised in the opening chapter.

Generations of biologists have banged their collective heads against the thigh of Venus in an attempt to untangle, at least theoretically, why sex continues. Already in 1889, German evolutionary theorist August Weismann, before the discovery of DNA, argued that sex could not have arisen simply to reproduce, since so many beings are capable of uniparental

reproduction. Rather, sex (which he called "amphimixis"), by mixing genes, increased the rate of genetic variation and therefore the chances of coming up with variants capable of survival.

Ronald A. Fisher emphasized that sex would tend to get rid of bad genes, because only half the chromosomes would end up in the parents' eggs and sperm. (The same would be true of "good genes," but they would tend to be preserved rather than eliminated by natural selection.) In uniparental organisms, deleterious genes would inevitably be handed down to offspring like hot potatoes in hell's kitchen, with no chance of reprieve through genetic shuffle. Meiotic sex—taking half of the chromosomes from the parent's body cells prior to reproduction, like a dealer cutting a deck in two prior to shuffling in Vegas—could hook you up. Compared with sexual organisms, one-parent organisms enjoyed fewer lottery tickets or pulls on the slot machine of success in evolution's genetic casino.

Scientists distinguish between the *origin* and the *maintenance* of sexuality. Ideas to explain the origins of sex—such as bacteria taking fragments of one another's genes on an ultraviolet-saturated primitive Earth—are not that easy to test experimentally. Scientists have thus tended to focus more on *why* sex exists. The orthodox answer is that, although sexual reproduction is more complicated than simple cell division or cloning, it confers genetic benefits on the many species of plants and animals that engage in it. By requiring the fusion of genes in each new individual, sex provides an opportunity to pool advantageous traits and mutations. Additionally, when deleterious mutations come together in unfit individuals, those individuals die, thus ridding the population of negative traits. No wonder they didn't have a chance. Following Fisher's lead, biologists argued that genetic *variety* is so useful that sexually reproducing organisms—who made sure to shuffle their genes each generation—would outcompete ones that didn't.

In 1973 Leigh Van Valen, an evolutionary biologist at the University of Chicago, put forward the "Red Queen" hypothesis, the forerunner of one of the most widely accepted theories of why sex exists. Van Valen's theory describes the way evolutionary change begets further change by altering the environment at large. The basic idea of the Red Queen is simple: that change begets change and that, if the living environment around you is changing, you had better, too, if you want to stay alive. The term comes from chapter 2 of Charles Dodgson's (writing under the pen name Lewis Carroll) 1872 *Through the Looking Glass,* where Alice dreams about a house in which everything is reversed, right-to-left, as in a mirror, and the chess pieces are alive. To see the garden she climbs up a hill, but it leads her back to the house. Picking up her pace to get there faster, she instead runs into the house. Talking flowers tell her that someone like her often comes through their garden. The person turns out to be the Red Queen, whom Alice seeks out but who recedes as she moves toward her; when she reverses directions, however, the queen is in her face, and they climb a hill, which the Red Queen says can be a valley. The key moment occurs at the top of the hill, where perplexed Alice, running as fast as she can to keep up with the queen, finds that neither is moving. "Now, here," explains the queen, "you see, it takes all the running *you* can do to keep in the same place."

Apart from a general presentiment of relativity paradoxes in physics and mathematics—which were explored by French mathematician Henri Poincaré and others before Einstein—the fable of the speedy runner who gets nowhere is particularly relevant to evolution. Unlike what many people still imagine (it is a bastard inheritance of our Newtonian and Cartesian intellectual background), the environment in which organisms live is not static, inert, or inanimate. Rather, the environment in which organisms live and, if possible, survive

to reproduce is itself suffused with living, evolving beings. What this suggests to the Red Queen theorists is that because the environment itself is evolving, sometimes organisms must do all the running (evolving) they can do just to keep in the same place (persist).

The principle has many recondite examples from the backwaters of snail biology, theoretical models of extinctions, and predator–prey ecology. But a simple example from the human realm will make the basic point clear. After the publication by Edward O. Thorpe of *Beat the Dealer* in 1966, blackjack became the most popular and most lucrative casino game, replacing craps. Thorpe, an MIT mathematician, showed that blackjack could be beaten by a card-counting system. Though difficult to implement, this radically increased the popularity of the game, making windfall profits for the casinos and card counters alike. Eventually, however, the casinos began to take (so to speak) countermeasures against the winners. They increased the number of decks, changed the rules (such as having the dealer hit on soft seventeen), introduced automated shuffling machines, and so on. Each casino corporation's new impediment to winning card play that was introduced spurred knowledgeable players on to find means around it. If the card counters had just stood idly by while the casinos implemented their changes, they might not have continued to win. They had to change their game, as the rules of the game changed, simply to maintain their winning ways.

This Red Queen dynamic is universal in mutually evolving systems. It applies to the sensory apparatus of predators, which must become more acute to get around the defenses of would-be prey evolving tougher hides, better camouflage, faster legs, and other escape mechanisms. It applies to weapons manufactures and war games, for example to the development of radar and radar cloaking devices in warfare. It applies to the interaction between flowering plants, which benefit by producing

sweet nectars, intriguing fragrances, and beautiful colors to attract animals to economically fertilize them and spread their seed, but which are also in danger of being eaten alive by those animals. The general type of response of the flowering plant is like that of the opinionated rose in the Looking Glass story who advises Alice to walk away from the Red Queen. Though not part of Carroll's conscious oeuvre, this rose, like all roses, attracts with petals and repels with thorns—an evolutionary example of a Red Queen response to botanical life in a world of disseminating but devouring animals. The great variety of chemical compounds produced by plants and fungi, with distinct effects on the bodies and minds of animals, are also to be interpreted in terms of plants evolving in a world of animals that by turns may be helpful or harmful. Red Queen dynamics also apply to mammalian intelligence in general and human society in particular, where there can be real advantages to knowing what somebody else is thinking (or plotting), as well as to conveying the impression that you are going to do or have been doing one thing when you have really been doing or are planning to do another (lying).

And the Red Queen process applies to sex, because organisms whose genes are continuously shuffled by meiotic sex once every generation make a moving, rather than a still target for the pathogens and parasites that might better infect and spread through their populations if they would only stay genetically put.

Evolutionary biologist Graham Bell of McGill University in Montreal has made use of the Red Queen hypothesis to explain the advantages of sexual reproduction. By shuffling their genes with each offspring, sexually reproducing organisms appear to be better at eluding parasites. Evidence comes from experiments with minnows—small freshwater fish—in Mexico. Researchers placed male-less populations of fish—they were spawned only from virgin mothers—and sexually reproducing minnows in

different ponds. They then counted black spots caused by a parasite, and found that the one-parent minnows were far more susceptible to disease than their genetically more varied, two-sexed relatives. By shuffling their genes every time they reproduce, sexual reproducers appear to be better able to elude the evolving bevy of potential inner assailants.

Hybrid vigor—the noted superior hardiness of organisms with genetically dissimilar parents—is also associated with beauty, at least in humans. The chances of facial and body symmetry that we find beautiful increase when separate genetic stocks are brought together. This phenomenon can also be seen from the inverse: Inbred fruit flies have less symmetry than those of more varied stock. And cheetahs, having little genetic diversity among remaining populations, tend to have asymmetrical facial bones.

Clear, disease-free skin is a key trait in human estimations of beauty. Thus, when lovers are attracted to each other, they may be unconsciously estimating each other's freedom from, or ability to ward off, potentially damaging parasites. Sexual recombination, mandating organisms to change and thereby elude many of their would-be parasites, amounted to a primeval form of preventive medicine.

Yet while much ink has been spilled over what "maintains" sex (why it continues to exist in sexually reproducing organisms), it's not as though we humans, and other plants, fungi, and animals, can just opt out of the process altogether. Even populations of all-female virgin minnows, lizards, and other animals develop from embryos and undergo cell fertilization-like processes, suggesting that they had two-parent ancestors. In most multicellular beings, sexual reproduction is the only way a new being can form. Sexual fusion of eggs via sperm starts the embryo, and all plants and animals develop from embryos. On paper, it is easy enough to say we could eliminate the "damnable expense" and just be cloned. However, given

the current state of science, and despite advances in assisted reproductive technologies, even the wealthiest misandrist (hater of men) cannot at present access the means necessary to completely bypass males if she wishes to reproduce.

A similar situation affects those all-female populations of animals officially classified as "asexual": They also have not completely circumvented their two-parent heritage. Careful study reveals that these organisms—such as the all-female populations of whiptail lizards that inhabit the southwestern United States—still undergo chromosome sex processes at the cellular level. Female whiptails mount one another and produce more eggs when they have been mounted by another female. One-parent animals appear to be evolutionary rarities that have devolved from two-parent animals rather than truly "asexual" beings. Otherwise why would they indulge in lesbian activities, mounting one another? Such all-female groups are not really species, since species consist of mating members. They are clones, uniparental populations. They are single parents, but they are still sexual, as their female frolicking suggests.

In order to understand the depths of sexuality, we don't just need to study animals and microbes, for the roots of sexual reproduction lie in events that are many hundreds of millions of years old. The Red Queen and other theories explaining why sex is genetically worth the trouble assume that most species of plants and animals could simply lose the complex of gender-determining and lover-performance traits of sexual reproduction if it were adaptive to do so. But this is not likely to be the case with such deeply entrenched traits, especially over the evolutionary short run.

We may wish we could fly by flapping our arms because it would be more "adaptive," but the physiological option is simply not open to us. Sexual reproduction may be so old that most species are unable to lose it. And those that do may become vulnerable to rapidly spreading parasites, and therefore removed

from the gene pool. Simon Robson, a biologist at James Cook University in Queensland, says of zoologists who theorize on the maintenance of sex without regard for its non-animal history: "They're working from a data set two billion years out of date."

But even if sex is a primordial genetic mixing mechanism selected for in evolution, this answers a decidedly secondary question. From a cosmic perspective, the tendency of all systems, as described by the second law of thermodynamics, is toward randomness. Chaos not order; disorganization not organization; dissolution not individuation. From a cosmic perspective, the question isn't why sexual organisms, but why organisms at all—organisms whose name shares roots with *organization,* a principle opposed to the way all material systems are supposed to work according to the second law. For the Greeks, the *organon* whence we derive the word *organism* was a tool, with a purpose. It was far from random. We, too, are far from random. We, and our organic brethren, sexual and non-, are organized. In fact, we can perhaps best be understood as natural tools, nature-made complex systems whose function is in service of the second law. Our organization, far from violating the second law, helps to illustrate how nature accomplishes its unconscious purpose, to come to equilibrium, spread energy, and promote atomic and molecular chaos—even if it means temporarily creating complex, materially cycling, energy-spreading structures to do so. The amazing thing about life isn't the differences it shows, which are exacerbated by mutation and sexual recombination, but its basic maintenance of its complex forms, implicit in the term *reproduction* but also notable in the maintenance of bodies known as metabolism, which maintains our and other creatures' energy levels and organization.

There would thus seem to be a conflict between evolution and thermodynamics. Indeed, after the defeats of creation-

ism in the US Supreme Court, religious campaigners reintroduced it as a scientific discipline under the name of Intelligent Design (ID). The campaigners, claiming the biological world is too complex to have been produced by biological variation and selection, argued that evolutionists are embarrassed by the second law of thermodynamics, characterized as the obvious tendency of nature to move from order and organization to disorder—to produce entropy.*

But, as the last sixty years of work in thermodynamics shows, there is absolutely no contradiction between the second law and the evolution of order and life. The second law applies to thermally sealed, so-called isolated systems, but the systems of life are open to material and energy flow. Not only life, but many sorts of naturally occurring complex systems spontaneously arise and grow in regions of energy flow.

Hyenas and anglers sure are mean. It's hard to imagine an all-powerful God with a love of fine craftsmanship using them to spread the light and moral exemplarity, let alone nonrecallable design. Grinding the bones of a striped zebra to white shit in the African grass is more easily understood in terms of natural selection for energy depletion. Complex systems feed on energy. We are complex systems whose ability to reproduce keeps the complex, entropy-producing process going. The passion of life, the fire in the loins, reflects a more general

---

* This refers to classical, thermodynamic entropy. *Entropy* is also a term used in communication or information theory. Since life obviously involves both information and energy, and because the two terms (which originated in a gambling equation) look almost identical in their respective equations, some have assumed they must be equivalent on some level. This is probably not true as, historically, *entropy* the word was suggested by mathematician John von Neumann to information theory founder Claude Shannon because, as he said, nobody knew what it meant anyway. This book talks about thermodynamic, not informational entropy. The natural tendency of energy to spread (acting like that's what it "wants" to do) appears to be the natural basis not only of life's existence as a complex system, but of biological urges such as hunger (directed at depleting concentrated chemical energy sources) and sex (directed at replacing the complex energy-depleting system before it wears out).

combustive process, one that sometimes goes to great lengths, and creates structured processes of dazzling intricacy—such as energy-dependent, energy-depleting human civilization—to accomplish its natural purpose.

Understanding why sex exists is for this reason a still fascinating but secondary question. The primary question is why we exist. We exist, in a basic, plebeian, profane, mundane, physical, and scientific if not ultimate religious and spiritual sense to perform a natural energy function. The reason is the same reason that tornadoes naturally spin into existence, maintaining, for a time, their complex systems—until the difference between high- and low-pressure systems is gone, at which point they also take their leave. Once all the agar on a petri plate is devoured, the bacteria consuming it die out. If all the energy of the sun could be turned into photosynthetic life, photosynthetic life would vanish. And if our evolving technological intelligence exploits all available sources of energy, we will also have a not-so-hot and perhaps premature date with fate.

Although sex is secondary to this cosmic process of energy depletion, on Earth it remains widespread. A rock-and-roll weekly that I saw in Boston in the 1980s took a poll as to the weirdest places one had had sex. One groupie wrote "in a meat packing plant." Another, "in the butt." But organisms also have sex in the bodies of species other than their own. In a strict biological sense, sex refers to the combining of genes from more than one source. By this definition, if a virus introduces stranger genes into your genome, even if you don't get sick, you have had sex. Not sex with a member of your own species, however. Viruses are not the only bearers of little pieces of nucleic acid that can merge with your own. Viruses are coated in protein, but plasmids are naked bits of DNA, taken in by bacteria, whose genes are not arranged into chromosomes but float relatively freely within the cell. Bacteria do not need sex to reproduce, as we saw in this book's beginning, but indulge nonetheless.

I call the attraction toward opposites, the tendency to join with others, even those of other species, xenotropia—from the Greek words tropos (turn) and xenos (stranger). Organisms turn toward strangers and also, in many cases, merge with them. Although they are not fully organisms, because they always need a live cell to make more of themselves, viruses, which we typically consider to be universally evil disease agents, biological if not diabolical pathogens, are in fact often beneficial. Long ago many of them turned toward and merged with animal genomes conferring what, in retrospect, turned out to be beneficial traits. Indeed, fully 8 percent of our genome appears to be remnants of infectious viruses, some of which may have been picked up in our primate lineage millions of years before we became human. There is evidence that some of these viruses were involved in the evolution of the brain as well as the immune system, in some cases even protecting us against other diseases. Viruses, because they pick up genes from one species and deliver them to another, sometimes transmit important survival information. Genetic mixing has been going on across species, let alone tribal, barriers for billions of years.

Recent research shows that the human placenta transcribes whole families of retroviruses, the same sort as are implicated in diseases, except that they are endogenous—meaning that they belong normally in the body and cause no ill effects. It thus appears that they are not only harmless, but actually required for placental and therefore fetal development. The famous cloned sheep Dolly died of a retrovirus. But experiments also show that interfering with the protein production of certain endogenous sheep retroviruses prevents the embryo from implanting on the placenta. A similar "reverse illness" from not having a given retrovirus in the body, or from its not functioning properly, may cause miscarriages in humans. The class of viruses in question, known as enJSRV (short for

"endogenous retroviruses related to Jaagsiekte sheep retrovirus"), is thought to have originally been infective, but then led to more efficient pregnancies in the small ruminants ancestral to sheep. The result is that genes picked up during infection—horizontal, bacterial-style sex—conferred evolutionary advantages.

Viruses are us: They became part of mammal genomes. Viral structural proteins can be "hijacked" and integrated into reproductive tissues, the immune system, and the brain. Some retroviruses disable receptors that lead to infection by other retroviruses. At bottom, we are part virus, the offspring not just of our parents but also of promiscuous, proliferating elements of DNA and RNA in our genome. Strangely, our very humanity depends on such ancient transgressions, no less than sheep would not be sheep without their acquired enJSRV.

We are also part bacteria. Genetically highly distinct types of bacteria merged into threesomes, foursomes, and moresomes. The results of some of these permanent matings were amebas and other cells with nuclei and chromosomes, including the ancestors to all animal cells. The most diehard puritan is walking testimony to viral promiscuity and bacterial orgies that, indefatigably, have persisted for billions of years.

But this is the "horizontal" sex of floating DNA fragments and interspecies alliances rather than the "vertical" sex of sperm and egg coming together to make offspring. This is the chromosome separation and fertilization of the DNA dance proper. This is reproductive sex, relatively recent in evolutionary terms, evolving only about a billion years ago.

The birds do it, the bees do it, the basidiomycote fungi* do it. But the bacteria don't, at least not to reproduce. Many

---

* These include *Schizophyllum commune*, an extremely successful mushroom found on every continent except Antarctica. It has twenty-eight thousand genders, each of which can mate with any of the others, except its own.

microscopic beings don't appear to engage in any version of the proverbial nasty, although they may still be getting it on behind researchers' backs. The naked amebas, the green swimming euglenids, the obscure bicosoecids with their double undulating arms appear to be sexless. The jakobids, a type of single cell with nuclei, or protist, just say no (as far as is known), and the bodonids, except for the intransigent *Dimastigella trypaniformis,* are confirmed bachelors of the protist world. Many fungi are chaste, and even those that do do the dirty deed, which often involves joining parental nuclei into a single cell, reproduce more often by asexual spores. Fragments of sponges break off, wafted about by water currents, growing into new individuals without any boinking or bonking. Flatworms undergo parthenogenesis, making new flatworms without putting on the submarine freak. The entoprocts, primitive animals with a weak muscular system and no heart, have a rudimentary nervous system consisting of a ganglion located above the digestive system with a few nerves extending to the stalk and tentacles. They do it sometimes. When they don't, they produce new individuals by budding. A few of their species are hermaphroditic, but fertilization itself has never been seen—although sperm with tails have been observed in their vicinity.

Of the thirty odd phyla of animals, most have members that can reproduce from one parent or two. Nonetheless, reproductive sex in some form, (for example at the cell level as self-fertilization and meiosis) appears to exist in all plants and animals, a great many fungi, and many organisms, such as slime molds and seaweeds belonging to that greatgroup (which is technically known as the protoctists of organisms that are not animals, fungi, bacteria, or plants). *Plasmodium,* the malaria-carrying parasite, gets it on only in the female *Anopheles* mosquito. The fertilized cell of *Plasmodium* transforms to a hard structure, which undergoes meiosis to produce infective cells, which migrate to

the salivary gland of the insect. Hijacking a ride on the mosquito bite, they are injected into the animal bloodstream, where they cozy up to red blood cells, feasting on the iron of hemoglobin before they transform into trophozoites, feeding cells that invade the blood cells and transform into a migrant form, merozoites, that reproduce in waves. After these waves (which affect us as chills and malarial fever), the merozoites differentiate into male and female gametes—who need to wait till they get back to their mosquito mother ship before they can join and start their whole complex sickening cycle over again.

Other protoctists rarely engage in sex. Whenever one species of Stentor does so, both lovers die. Their demise and final disintegration takes four days, but it is the inevitable consequence of entering the mating act. Romeo and Juliet never have offspring. Sex is a dead end for this species.

The great variety of cell sex in the vast kingdom of protoctists suggests that our kind of sex began in them. Darwin and Nietzsche both warned that we should not confuse origin with present function. The DNA dance may make sexual reproducers better at eluding pathogens than organisms with only one parent, which don't have to go through sex to reproduce. But being born of one rather than two parents is not the same thing as being free and clear of all cellular sexual processes. Meiotic sex, by which we lose ourselves in the lives of others, is not that easy to lose. If it is not possible to lose all aspects of meiotic sex in animals, the answerable question is not, Why is sex maintained? but, How did it originate? It is to this intriguing question that we turn next.

Silence of the Amebas

*W*hat does Hannibal Lecter, the eponymous cannibal of film fame, played by actor Anthony Hopkins, have to do with the origins of sex? On the surface, not a helluva lot, although I did find myself, once upon a time, at the entrance to the park on Riverside Drive in Manhattan, reading a sign that said STAY AWAY FROM THE RACCOONS, THEY MAY HAVE RABIES, when a small compact man, whose face looked incredibly similar to that of Anthony Hopkins (only it was nowhere near as large as it appeared on the silver screen), scoffed and, in a strangely clipped accent that combined vaudeville Hollywood with a tone of British authority, proclaimed that rabies had nothing to do with the daylight sightings of slow-moving raccoons such as the fat ones we were both observing right now. "They are pregnant," he huffed. "Raccoons always come out in the daytime when they're pregnant. People," he added, in supercilious tones that transcended mere playacting, "are the most ignorant species on the face of the Earth."

Dumbfounded and amused, I hesitated even to agree with him (which I probably would have), harboring at the same time the sneaking suspicion that he, despite his imposing delivery, was talking through his hat. At which point a long black limo pulled up alongside us and the head of a middle-aged blond woman, who looked remarkably like Martha Stewart, appeared over a descending window. As the confident stranger turned to the car, I noticed a handkerchief sticking out of his back pocket. Such are our brushes with greatness.

Although he only appears in the film for some sixteen minutes, Sir Anthony as Hannibal Lecter was voted by the American Film Institute to have been the most memorable villain in film history. The author of the novel series on which the movie is based, Thomas Harris, portrays Hannibal the Cannibal's backstory in the 2006 prequel *Hannibal Rising*. The extremely intelligent psychopath, noble Lithuanian and Italian blood running in his veins, got off on the wrong track or, as they say, out of the wrong side of the bed, human-wise, when, in 1944, German and Soviet forces ransack the family estate and, later, deserting members of the Einsatzgruppen, the Nazi "task forces" charged with killing Jews, Gypsies, and Soviet military, murder and then eat Hannibal's beloved sister Mischa before his horrified eyes. The character then goes on to avenge his sister's cannibalism, tracking down, torturing, and killing his sister's assailants one by one.

Although cannibalism has been practiced ritually by some cultures, primarily indigenous peoples with a custom of eating their enemies, it violates a deep social taboo. If I, a hungry person, can eat you, then by rights you, a hungry person, should be able to eat me. This puts everybody in jeopardy. I can eat a hamburger or a steak or a chicken McNugget because, after all, the animal who gave up its body to feed me is not like me. Except for vanishingly rare occasions—such as the eye-pecking birds in Alfred Hitchcock's horror classic *The Birds*—I don't have to worry about the tables being turned. The word *cannibal* itself derives from the Carib word *karibna,* meaning "person" in the native language of the Carib or Kalinago people of the Lesser Antilles islands; the existence of tribal ancestor bones in their houses may have been the source of rumors of their cannibalism, rumors kept alive by Queen Isabella's 1503 ruling that cannibals, as it would in her view still be a step up for them, could still be taken as slaves.

Ritualistic and psychopathological reasons aside, the main

evolutionary reason for cannibalism is clear. Hunger, when it reaches an acute state, leads organisms to eat their own. Thousands of species, including lions, chimpanzees, and bears, will eat their own kind. Pigs, living up to their name, often devour piglets. Sharks, still worse, sometimes devour their unborn sibs before either is born. Even herbivores occasionally indulge. Praying mantis and black widow spider females eat their mates after copulation. Insects, with their obvious lack of moral strictures, are among the worst offenders. In some species, males make a spermatotheca—a nutritional gift package containing sperm that the female takes into her body. Apparently no big intellectual leap is needed for her to take the next step.

All this rampant eating and devouring of others and one's own relates directly to a likely scenario for the origins of our kind of meiotic sex. In fact, in the 1940s Harvard University biologist Lemuel Roscoe Cleveland solved the mystery of the origin of fertilization and how cells with one set of chromosomes (haploid cells) came to merge with each other and develop two sets of chromosomes (becoming diploid). Cleveland noticed that in some cases, hungry ameba-like cells would incompletely devour one another, merging cell membranes and cytoplasm and pooling their chromosomes in a single nucleus. He reasoned that the first fertilization was the result of not a sexual but a starvational urge to merge.

This theory is borne out by evidence from other quarters. Some cells today reproduce asexually . . . at least until they are faced with starvation, such as from lack of nitrogen in their environment. In this case, they fuse together to make a diploid. Cleveland witnessed cannibalism among microbes called hypermastigotes as well. They merged cell membranes and nuclei. Without a stomach or an immune system, it's easier for your food to live inside you, like Jonah in the whale.

Cleveland also carefully chronicled and filmed the necessary separation from the state of doubleness, which he called

"relief from diploidy." We undergo such relief when we produce haploid sperm cells or ova from diploid body cells; aside from these sex cells, the rest of the cells in our body remain in a diploid state, with forty-six chromosomes, twenty-three each from Mum and Dad.

The evolution of sex, let alone its maintenance, is not one thing. Although bacteria recombine genes, the evolution of reproductive sex occurred in these silent beings, the protists, single cells with nuclei and chromosomes. Beyond the theoretical advantages of recombination we must look to these wildly diverse ameboids to understand the peculiar cycles and attractions that all but define our lives. Cycles of hungry doubling and diploidy relief—the origin of reproductive sex—are thought to have arisen at least three times in the history of life: in the choanomastigotes ancestral to animals, the chytrid cells ancestral to fungi, and the green algae ancestral to plants. Some, once they merged, reproduced to make bodies. Each body cell had chromosomes from both parents. But to develop new bodies cells had to return to their ancestral state, the so-called haploid state of sperm and eggs with only one set of chromosomes each. Indeed, to get back to the stem cells from which bodies and tissues can be grown from the fertilized egg, "we" must die.

It sounds like a line from Hannibal Lecter: "As a sexual being, Agent Starling [the Jodie Foster character in the movie], 'we' must die."

Creepy, but true.

Meiosis reduces chromosome numbers. Fertilization routinely doubles them. But so, accidentally, did cannibalism. I didn't get a chance to tell Hopkins, but sex seems to have had a most unromantic genesis.

Bliss

*A*lone on a beach in Saint Martin in the Caribbean, find-ing solace by the ocean after a fight prior to the final passionate sunset of a relationship that was forever ending, I saw a sun-streaked scene of such poignant tranquility that I may never forget it—and even if I do I will still, in a way, remember it, as the scene itself was one of sweet bodily surrender and memory abandon.

I don't remember if I was wearing clothes, but the couple I was watching weren't. Having trekked past vendors selling tropical shirts, tropical drinks, and subtropical massages, I had moved closer to the shore and to the farthest end of the beach marked by metal construction. There before me, wading into the water, were an old man and a woman. He cradled her in the shallow ocean. The motions were of such exquisite tenderness I could not help but watch. The man's devotion was complete; the meaning seemed clear: Here was the love of his life, on its last legs. But that no longer mattered: They were together in an eternal moment, the bliss of a love that would never die. Together, as a single form, they were bathed by crystalline blue waves. The woman, completely relaxed, leaned back, as if to merge with the sun.

Without sex, there would be no holding of hands, no pressing of lips. There would be no tears of betrayal, no erotic dreams. There would be neither bouquets (made of flowers, plant sex organs) nor wine (the fermented fruit of the grapevine, a flow-ering plant). There would be no women. There would be no birds. And there would be no love.

It would be enough if the sexual urge gave us love and death. But it does more. Sex relieves tension, goes the line; love creates it. But sexual attraction ensures that organisms will be linked in many ways, not just sexually. The energy that would go into the chase in earlier times helps organize individuals into complex aggregates, and the ancient energy of eros, redirected and domesticated, drives more modern pursuits, such as art, work, and purchasing of products whose advertisers lure us with by turns subtle and blatant sexual imagery. Sexual energy is more than enough to get the job of reproduction done. Its excess spills over into nonreproductive pursuits, including what Wilde called "the love that dare not speak its name" (homosexuality). The pleasures, excitement, and energy of the redirected sex urge show up in chaste flirtations and lifelong friendships. Lesbians and homosexuals enjoy sexual pleasure, Mary Roach reports in her recent book *Bonk,* more than their more heterosexual counterparts. If so, the enjoyment that in part drives us to repeat the acts that lead to reproduction would appear to be more than sufficient to get the job done.

The urge to merge does not stop demurely at its "proper" objects but continues blundering on. The foot fetishist is smitten with the "wrong" part of the body in that fertilizable eggs are located a few curves away in another part of the female anatomy. So, too, the fetish for rubber, which, not live tissue (although consisting of organic carbon-hydrogen compounds), stops sperm dead in their tracks. Such "mistakes" can be understood as the untoward impressions made upon young animals at an early stage of development. Some young gulls, literally gullible, if exposed early to the distinct eyes of another species, attempt later to mate with its members. Gullible to the end, they never reproduce. Similarly, if a toddler is exposed at a crucial point in his life to the sight and smells of triangular boot leather, he may get the wrong idea. Nonetheless the species itself lurches sufficiently vulvaward to preserve.

The diverted urge to merge also has evolutionary consequences. Consorting among radical strangers is not confined to the subvisible realm. Donald Williamson from the Isle of Man quips that he is "from a short-lived family but on a straight line course for posthumous recognition." A kind of modern-day Dr. Moreau, Williamson successfully used the sperm of a male sea urchin, *Psammechinus miliaris,* to fertilize eggs of the sea squirt, *Ascidiella aspersa*. With his help the marine invertebrates did it across not just species but phylum boundaries—as if a man mated not a seal or a horse (both in our phylum) but a squid or a sponge.

Nonetheless, the union was fertile, with offspring in the form of four-armed pluteus larvae, captured in computerized photomicrographs by researchers from Holland. The Dutch researchers independently verified Williamson's hybrid, which also budded into spheroids, suggesting that the new cross-species life-form could reproduce. The slender, personable Williamson, now confined to a wheelchair, draws our attention to metamorphosing marine and insect species whose larval forms not only strikingly differ from the adult stages, but also resemble the adult forms of other species. Williamson's brainchild is that metamorphosis, both insect and marine, reflects primordial hybridization events. If, Williamson reasoned, the adult forms of metamorphosing insects and marine invertebrates had been sired by different species—an easier occurrence when eggs are fertilized outside the body—then would-be transgressive expressions of the sex urge may not have always been entirely evolutionarily untoward. Some of the most beautiful beings on the planet may ultimately have parents from different species. Their parents—stupid or perverted or at least inaccurate enough not to fertilize the "right" mate—led to bursts of evolutionary novelty. The change from caterpillar to butterfly, and pluteus larva to starfish, may be so dramatic because two genomes were evolutionarily integrated into one.

Indeed, Williamson speculates that the strange animals of the Burgess Shale, and the original explosion of animals in the fossil record of the Cambrian era, may owe not a little to such ancient transgressions.

The transgressive tree of life is really a bawdy bush. Its branches not only diverge but merge back into one another. This process should not be underestimated. Some argue that both AIDS and cancer will only be truly understood once we understand the symbiotic, parasexual cell history of the mitochondria whose role in the immune system, where they are involved in intracellularly deploying oxygen- and nitrogen-containing gases to fight pathogens, was only made in the 1990s, after the lucrative AIDS infrastructure became entrenched. Although they have been, like sex slaves, parasexually trapped in the gilded dungeon of the eukaryotic cytoplasm for some two billion years, the mitochondria still have their own DNA, their own reproductive timetable, and their own needs. If these needs—particularly for oxygen—are not met, cancer and AIDS may be the result. Despite the billions that have flowed into medical foundations, research institutions, insurance companies, and government programs, there is growing evidence that AIDS is caused not by a virus (many people with AIDS lack HIV and vice versa) but by mitochondrial malfunction. Our deep nature is not unitary. It is multicellular, sexual, parasexual, and symbiotic. We are not pure individuals but mixed beings, dependent upon further genetic mixing to keep going. We are not medical abstractions but xenotropes, attracted by and attracted to others.

A Tibetan mystic saying goes: We are here to realize the illusion of our separateness. The spiritual sentiment has a biological cognate. Our xenotropic drive—to merge with what is not us, temporarily in sex, or permanently in symbiosis or cross-species hybrids—is more than a metaphor. But it also offers spiritual solace. When we hook up with another, in sex

or love (or, more rarely, both) we prove that our isolation is not permanent. In the fullness of time, we may all be linked. In the meantime, eros brings us together, making us more than we are alone. Cupid's arrow, quivering into the heart of loneliness, kills us even as it sets us free.

FURTHER READING

Bataille, Georges. *Visions of Excess: Selected Writings, 1927–1939.*
   Minneapolis: University of Minnesota Press, 1985.

Bataille, Georges. *Story of the Eye.* New York: Berkley Books, 1977
   (1928).

Bell, Graham. *Sex and Death in Protozoa: The History of Obsession.*
   Cambridge: Cambridge University Press, 2008.

Bukowski, Charles. *Erections, Ejaculations, Exhibitions and General Tales of
   Ordinary Madness.* San Francisco: City Light Books, 1967.

Klossowski, Pierre. *Sade My Neighbor,* translated by Alphonso Lingis.
   Evanston, Illinois: Northwestern University Press, 1991 (1947).

Margulis, Lynn, and Michael J. Chapman. *Kingdoms and Domains: An
   Illustrated Guide to the Phyla of Life on Earth.* London: Academic Press,
   2009.

Margulis, Lynn, and Dorion Sagan. *Origins of Sex: Three Billion Years
   of Genetic Recombination.* New Haven, Conn.: Yale University Press,
   1990.

Morgan, Elaine. *Aquatic Ape Hypothesis.* London: Souvenir Press, 1999.

Reage, Pauline. *Story of O.* Philadelphia: Running Press, 1998.

Ridley, Matt. *The Red Queen: Sex and the Evolution of Human Nature.* New
   York: Harper, 2003.

Sade, Marquis de. *Justine, Philosophy in the Bedroom, and Other Writings.*
   New York: Grove Press, 1990.

Schneider, Eric D., and D. Sagan. *Into the Cool: Energy Flow,
   Thermodynamics, and Life.* Chicago: University of Chicago Press, 2005.

Smith, Robert L., editor. *Sperm Competition and the Evolution of Animal
   Mating Systems.* London: Academic Press, 1984.

Vernadsky, Vladimir. *The Biosphere: Complete Annotated Edition.* New
   York: Springer-Verlag, 1998 (1929).

REFERENCES

Bettie Page friend on nudity versus nakedness: Cited in Dargis, Manohla, "Always Comfortable in Her Own Skin [Bettie Paige Appraisal]," *New York Times,* Saturday, December 13, 2008, C1.

Primate research: Primate Info Net: Library and Information Service, National Primate Research Center, University of Wisconsin–Madison: http://pin.primate.wisc.edu/aboutp/evol/index.html.

Hair protein: Winter, H., L. Langbein, M. Krawczak, D. N. Cooper, L. F. J. Suarez, M. A. Rogers, S. Praetzel, P. J. Heidt, and J. Schweizer. "Human Type I Hair Keratin Pseudogene phihHaA Has Functional Orthologs in the Chimpanzee and Gorilla: Evidence for Recent Inactivation of the Human Gene After the *Pan–Homo* Divergence." *Human Genetics* 108, no. 1 (2001): 37–42.

Homo paniscus: Hecht, Jeff. "Chimps Are Human, Gene Study Implies." *New Scientist:* www.newscientist.com/article/dn3744-chimps-are-human-gene-study-implies.html (May 19, 2003).

Sperm competition: *Sperm Competition in Humans: Classic and Contemporary Readings,* edited by Todd K. Shackelford and Nicholas Pound. Springer, 2006.

Human intelligence as the result of sexual selection by females "shopping" for interesting genes: Miller, Geoffrey F. *The Mating Mind: How Sexual Choice Shaped the Evolution of Human Nature.* Anchor Books, 2001.

Humans as a "musical primate": Vaneechoutte, Mario, and John R. Skoyles. "The Memetic Origin of Language: Modern Humans as Musical Primates": http://cfpm.org/jom-emit/1998/vol2/vaneechoutte_m&skoyles_jr.html#sect6.1.

Giordano Bruno: Cited in Krumbein, Wolfgang E., and Betsey Dexter Dyer. "This Planet Is Alive: Weathering and Biology, a Multi-Faceted Problem," in *The Chemistry of Weathering,* edited by J. I. Drever. Boston: D. Reidel Publishing, 1985, 145.

Bataille epigraph: From the introduction to *Eroticism,* translated by Mary Dalwood. London and New York: Marion Boyars, 1962 (1957): http://home.wlv.ac.uk/~fa1871/battext.html.

Sade quotes on morality: www.brainyquote.com/quotes/authors/m/marquis_de_sade.html.

Incest in *Adactylidium:* "Death Before Birth, or a Mite's *Nunc Dimittis,*" in Gould, Stephen Jay. *The Panda's Thumb: More Reflections in Natural History.* New York: W. W. Norton, 1980, 69–75.

"I caused the Montreuils . . ." Sade, cited in Neil Schaeffer, 2001. "Perverting de Sade," *The Guardian,* Saturday January 13: www.neilschaeffer.com/sade/bibliography/quills.htm.

Sacher-Masoch, Leopold von. *Venus in Furs,* translated from the German by Fernanda Savage, 1921: http://books.google.com/books?id=W12 DWT2jW6UC&printsec=frontcover&source=gbs_similarbooks_r&c ad=4_2#PPA88,M1, 88.

Retroviral bacterial-style sex preventing miscarriages in sheep: "Researchers Discover That Sheep Need Retroviruses for Reproduction," *ScienceDaily,* September 11, 2006: www.sciencedaily .com/releases/2006/09/060911233630.htm.

Bulwer-Lytton writing contest: www.bulwer-lytton.com.

Bulwer-Lytton Fiction Contest 2008 results: www.sjsu.edu/faculty/ scott.rice/blfc2008.htm.

Meadows, Robin. "Sex and the Spotted Hyena." *ZooGoer,* May–June 1995: http://nationalzoo.si.edu/Publications/ZooGoer/1995/3/ sexandthespottedhyena.cfm.

Why we have sex: Sherratt, Thomas N., and David M. Wilkinson. *Big Questions in Ecology and Evolution.* Oxford University Press, 2009. See also Sagan, D. "A Brief History of Sex." *Cosmos* (Australia), June–July 2007, 50–55.

"He was still watching . . .": Nin, Anaïs. *Delta of Venus: Erotica.* Harvest Books, 2004, 77.

Intrauterine testosterone and ring fingers: See, for example, Smith, Adam. "Successful Traders: The Testosterone Effect." January 12, 2009: www.time.com/time/business/article/0,8599,1871066,00 .html?iid=tsmodule.

Red Queen and blackjack: Thorpe, Edward O. *Beat the Dealer: The Book That Made Las Vegas Change the Rules.* New York: Random House, 1966.

Chimpanzee human-like behaviors: Waal, Frans de. "Obviously, Says the Monkey." In John Templeton Foundation, *Does Evolution Explain Human Nature: Twelve Views on the Question,* 2009, 4–7: www.templeton.org/evolution. And see Waal, Frans de. *Primates and Philosophers: How Morality Evolved.* Princeton, NJ: Princeton University Press, 2009.

". . . the traces of my resting place . . . minds of men": Encylopaedia Britannica, "Marquis de Sade": www.britannica.com/EBchecked/ topic/515876/Marquis-de-Sade/6335/Writings.

"The individual which may be born . . . meditation": Schopenhauer, Arthur. "Metaphysics of Love." In *Essays of Schopenhauer.* ebooks@ Adelaide, 2004: http://ebooks.adelaide.edu.au/s/schopenhauer/ arthur/essays/chapter12.html.

INDEX

Index

Index

ABOUT THE AUTHOR

Dorion Sagan is the author of numerous articles and books translated into eleven languages, including *Notes from the Holocene: A Brief History of the Future*; *Into the Cool: Energy Flow, Thermodynamics, and Life* (with Eric D. Schneider); and *Up from Dragons: Evolution of Human Intelligence* (with John Skoyles). His *What Is Sex?* (with Lynn Margulis) was chosen as one of five "mind-altering masterpieces" by *Utne Reader*. Sagan's essays are included in collections edited by Richard Dawkins and E.O. Wilson. He graduated from the University of Massachusetts—Amherst with a degree in history and has wide-ranging interests in philosophy and literature.

Reviewing Sagan's *Microcosmos* in *The New York Times Book Review*, Melvin Konner wrote: "This admiring reader of Carl Sagan, Lewis Thomas, and Stephen Jay Gould has seldom, if ever, seen such a luminous prose style in work of this kind." Sagan has written for *The New York Times*, *The New York Times Book Review*, *Wired*, *Skeptical Inquirer*, *Pabular*, *Smithsonian*, *The Ecologist*, *Omni*, *Natural History*, and many other publications.